# Corals - Habitat Formers in the Anthropocene

*Edited by Giovanni Chimienti*

Published in London, United Kingdom

Corals - Habitat Formers in the Anthropocene
http://dx.doi.org/10.5772/intechopen.98140
Edited by Giovanni Chimienti

Contributors
Henrique F. Santos, Luanny Fernandes, Flávia L. Carmo, Hugo E. de Jesus, Adriana U. Soriano, Walter Dellisanti, Davide Seveso, James Kar-Hei Fang, Íris Sampaio, Lydia Beuck, André Freiwald, Gui M. Menezes, Thomas J.F. Goreau, Pasita Chaijaroen



First published in London, United Kingdom, 2023 by IntechOpen
IntechOpen is the global imprint of INTECHOPEN LIMITED, registered in England and Wales, registration number: 11086078, 5 Princes Gate Court, London, SW7 2QJ, United Kingdom

British Library Cataloguing-in-Publication Data
A catalogue record for this book is available from the British Library

Additional hard and PDF copies can be obtained from orders@intechopen.com

Corals - Habitat Formers in the Anthropocene
Edited by Giovanni Chimienti
p. cm.
Print ISBN 978-1-80355-807-3
Online ISBN 978-1-80355-808-0
eBook (PDF) ISBN 978-1-80355-809-7

# Meet the editor

Dr. Giovanni Chimienti is a marine biologist at the University of Bari and at the National Interuniversity Consortium for Marine Sciences (Italy). He is also a member of the IUCN World Commission on Protected Areas and National Geographic Explorer. His main scientific research is about corals and benthic habitats, with particular reference to taxonomy (order Antipatharia, Alcyonacea and Pennatulacea), biodiversity, and population structures. He uses submersibles and remotely operated vehicles for deep-sea explorations, as well as data collected during fishing operations. He also studies the effects of anthropogenic impacts on coral populations. In the course of his work, Dr. Chimienti has discovered some spectacular mesophotic and deep coral habitats in different areas of the world, described new species and participated in the process of protecting some of the most strategic coral habitats in the Mediterranean Sea. Dr. Chimienti has received several prestigious scientific awards for his research, including from Accademia Nazionale dei Lincei, the Mediterranean Science Commission, the Deep-Sea Biology Society, and Prince Albert II of Monaco.

# Contents

# Preface

Corals play a key structural role in the creation of marine habitats ranging from the shallow, euphotic zone to the deep, aphotic one. Stony corals can structure coral reefs and frameworks, arborescent octocorals and antipatharians can create coral forests, and pennatulaceans can form large aggregations on soft bottoms known as sea-pen fields. All these coral habitats host a rich associated community, support unique symbiotic relationships, and provide a suite of ecosystem goods and services. This book provides updates, new findings and insights for an increased understanding of corals *sensu lato*, with a look at how humans influence the distribution and the conservation status of corals.

The book starts with an overview of the nutrition of corals and their trophic plasticity under future environmental conditions. Food availability is one of the main factors driving corals' occurrence and distribution, combined with the symbiotic relationship with photosynthetic endosymbionts, bacteria, and viruses. This delicate balance is now threatened by both direct and indirect anthropogenic impacts, including climate change, but some coral species have found a way to cope with fluctuating conditions and modulate their energetic budget. Unlike their shallow counterparts, deep-sea corals do not rely on photosynthetic symbionts, and their distribution is mostly shaped by substrate availability and by marine currents that carry nutrients and oxygen. Chapter 2 reports a case study off Mauritania (NE Atlantic Ocean) where deep-sea explorations have highlighted the presence of spectacular coral forests locally dominated by octocorals in the genus *Swiftia*. This habitat is a vulnerable marine ecosystem of high conservation importance, and many associated species may depend on it to feed, hide, or reproduce. The Timiris Mound Complex is particularly important for the presence of *Swiftia* forests, but fishery and deep-sea exploration targeting hydrocarbon resources occur in the nearby area. Sedimentation plumes caused by both activities can affect settlement, development and survival of these vulnerable corals and their associated community to a remarkable degree. Hence, the main features of these deep-sea biodiversity hotspots should be considered in conservation planning and design for protected marine areas.

Alongside fishery and hydrocarbon search/extraction, the unfortunate impact of oil spills on coral habitats is among the effects of human actions on the natural capital. Chapter 3 illustrates the effects of oil spills on coral reefs, the major historical records of oil spills and the possible remediation actions available today. Oil spills occur worldwide, from areas characterized by direct oil extraction activities to almost pristine places, as the main oil transport routes in the oceans are close to important coral reefs. Unfortunately, there is still no efficient, sustainable and large-scale remediation strategy to protect or clean up reefs impacted by oil spills. Furthermore, the chemical dispersants that are often used to remediate oil pollution in marine environments can also have negative effects on the living corals and their associated community. Efforts to reduce the consumption of fossil fuels and to find better solutions to prevent oil spills are complemented by emerging scientific scenarios that include the potential manipulation of the coral microbiome to mitigate the impacts of oil spills on coral reefs.

Efforts and resources for marine remediation are one of the main challenges in the Anthropocene, although active conservation, for example through protected marine areas, and drastic prevention of human impacts through appropriate marine policy, remain the most sustainable and effective ways to protect marine ecosystems, including coral habitats. Once damage occurs, however, habitat restoration practices can help to mitigate the negative effects on the biota. One of the effective ways to restore coral reefs affected by climate change-related anthropogenic impacts has been the use of artificial platforms that mimic the coral reef and provide a suitable substrate for several benthic species. The new generation of underwater structures, illustrated in Chapter 4, uses seawater electrolysis to create growing limestone frameworks to repair or replace damaged coral reefs.

Although remediation can be effective in mitigating the effects of human impacts, enforcement of conservation measures remains the priority to protect coral habitats, as the ecosystem goods and services provided by natural, wild habitats are more valuable than those provided by restored ones. Chapter 5 explores human–coral interactions from a socioeconomic point of view, highlighting how coral reefs benefit humans and how their deterioration affects the well-being of local populations. Many human communities on the coast rely directly on coral reefs for food, tourism income and coastal protection. At a local scale, the socioeconomic impact of coral reef degradation on coastal communities is illustrated through a case study of the 1998 coral bleaching in Indonesia.

The five chapters presented here deal with some of the more current aspects of corals and their vulnerability to human actions, adding a few more tiles to the still incomplete mosaic of knowledge about corals from tropical to polar latitudes, from shallow reefs to the unseen depths of the ocean. Appropriate and specific policy interventions are needed to preserve coral habitats, from the shallow to the deep, and as a consequence, to protect all those ecosystem goods and services from which all humankind benefits. In planning conservation measures, it is important to keep in mind that we do not manage the environment but we can manage human activities, trying to reduce our footprint as much as possible and to mitigate the negative effects of certain actions. This is going to be one of the greatest challenges of our time. Scientific research can produce the knowledge to understand, and support decision-makers in choosing the right path. In that regard, the tireless work of many marine scientists and conservationists, some of whom have contributed to this book, is warmly acknowledged. I am also grateful to Nera Butigan of IntechOpen for the efficiency and patience she showed while following every phase of this book.

**Giovanni Chimienti**
Department of Biosciences,
Biotechnologies and Environment,
University of Bari Aldo Moro,
Bari, Italy

Chapter 1

# Nutrition of Corals and Their Trophic Plasticity under Future Environmental Conditions

*Walter Dellisanti, Davide Seveso and James Kar-Hei Fang*

## Abstract

Scleractinian corals obtain metabolic energy from their endosymbiotic autotrophic microalgae, and from remineralization of organic matter by bacteria and viruses, along with the heterotrophic food sources. The mutualistic symbiosis is generally stable but can be disrupted when environmental conditions surrounding the corals, such as increasing seawater temperature, become unfavorable to sustain each component of the holobiont. In this connection, the effects of global stressors such as climate change, and local stressors such as pollution, and their combination, are posing serious threats to the metabolic resistance of corals. However, some more resilient coral species have developed specific mechanisms to cope with fluctuating environmental conditions according to the trophic strategy (autotrophy, heterotrophy, or mixotrophy), and by modulating their energy expenditure. In this chapter, the role of nutrition in the coral symbiosis as the energetic budget for metabolic performance will be discussed, with a focus on the role of acquisition of nutrients through feeding, regulation of energy reserves (lipids, proteins, and carbohydrates), and adaptation capability in the natural environment, including the expression of heat-shock proteins (Hsps). Future environmental conditions under a combination of global changes and local impacts will also be discussed, with the aim of identifying the trophic niches of corals and geographical areas as possible refugia.

**Keywords:** energy, metabolism, adaptation, resistance, climate change

## 1. Introduction

Scleractinian corals are complex key habitat-forming organisms that create biogenic reef structures from shallow to deep waters [1], and they are fundamental to the supporting of the biodiversity of the world's oceans. They have evolved to thrive in conditions of optimal nutrient availability [2], seawater temperature, and oxygenation [3], and are in competition for space with other benthic taxa [4]. Mostly distributed along the tropics, corals can be found also in high-latitude subtropical areas and deep seas [5–8], where they show adaptive capability to live in fluctuating environmental conditions [9]. Corals can develop several biological structures depending on their capacity to grow via vertical or horizontal extension (**Figure 1**). These biogenic habitats formed by coral reefs represent one of the worldwide hotspots of biodiversity

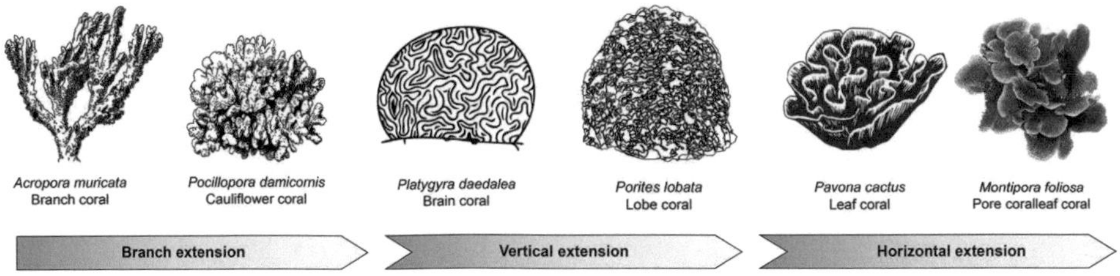


**Figure 1.**
*Major scleractinian coral morphologies and different colony growth modes, based on branch, vertical, and horizontal extension. Modified from Pratchett et al. [10].*

in the ocean [11], hosting a great variety of organisms, such as fish, macroalgae, and microorganisms [12].

Coral reefs can provide ecosystem goods and services, such as the provision of food, touristic activities, and protection of coastline from flooding and tidal movements [12, 13]. However, in the era of Anthropocene, coral reefs are among the habitats on Earth that are suffering the most and are dramatically degrading, since a multitude of factors are plaguing these marine ecosystems. Abiotic factors such as abnormally elevated or reduced temperatures, ocean acidification, high ultraviolet radiations, and fluctuation in salinity are increasing the occurrence of coral bleaching events [14–18]. Additionally, industrial pollution, coastal development, and excessive nutrient input, as well as biotic stressors such as predation outbreaks, epizootic diseases, and bioerosion are leading to further coral reef degradation around the world [19–22]. For all these reasons, corals are sensitive to changes in environmental conditions and therefore are considered good bioindicators of the health status of the marine environment [23, 24].

Corals are considered meta-organisms because of the complex biological interactions between the animal host and endosymbionts. Indeed, the concept of corals as holobiont encompasses the symbiotic relationship between dinoflagellate endosymbionts (Symbiodiniaceae [25]) and the animal host tissue (coral polyps), as well as the associated microorganisms found in coral tissue, gastric cavity, and coral skeleton. All components contribute to coral growth through the combined uptake of inorganic nutrients and food particles, photosynthesis, and deposition of calcium carbonate. In particular, the symbiotic relationship of corals is a mutual relationship between the coral polyps and the dinoflagellate endosymbionts. To gain metabolic energy, scleractinian corals are able to shift from heterotrophy (catching particulate food) [26, 27] to autotrophy (through photosynthesis by endosymbionts) [28]. Depending on the species-specific trophic strategy [29, 30], corals exhibit the ability to collect food particles (e.g. zooplankton) as a heterotrophic source of energy. On the other hand, they can rely on the autotrophic system of endosymbionts as an alternative source of oxygen and carbohydrates for aerobic respiration [31]. Oxygen availability determines the balance between the aerobic and anaerobic metabolic pathways, and therefore has significant implications for the energy budgets of corals [32]. These processes, however, are not perfectly balanced. Some species rely more on heterotrophy as an external source of energy, but some more on the photosynthetic system, while others are mixotrophic, meaning that they can increase the ability to modulate energy availability depending on the environmental conditions [30]. In all cases, the energy produced during the metabolic processes, which is stored as adenosine triphosphate (ATP), is used for maintaining the cellular physiology and supporting the intracellular uptake of dissolved inorganic carbon to form calcium carbonate, which is necessary for building the skeleton and sustaining the growth of corals [33]. Energy reserves include proteins, lipids, and carbohydrates [27, 34] can be used when there is a high energy demand, e.g. under thermal stress [35].

Corals also harbor a large variety of microorganisms on their surface, which contribute to biogeochemical cycles and the provision of micronutrients. For instance, bacteria, archaea, and viruses play fundamental roles in the remineralization of organic matter into micronutrients [36]. The nutrition of corals is linked to the uptake of macro- and micronutrients that support the metabolic processes and growth [26]. The roles of micronutrients, such as nitrogen and iron, in enhancing the capacity of symbiosis have also been highlighted, in particular for the endosymbionts to resist abnormal conditions of surrounding waters [34]. The microorganisms living on the coral surface and in the tissue are also related to the probiotic diversity necessary for the general health of corals [37]. In case of disruption of the symbiotic equilibrium during extreme events (e.g. heatwaves or nutrient discharge) and prolonged disturbances (e.g. climate change or pollution), the microbial community can change from the symbiotic to commensal mode, a change that could reduce the capacity of the coral host to maintain the metabolic equilibrium [38].

In this context, the coral holobiont is capable of gaining metabolic energy from multiple sources and therefore has the capacity to modulate its physiology depending on nutrient availability and environmental conditions. The continuous pressures from anthropogenic activities are leading to substantial changes in the capability of corals to develop resistance mechanisms, which in turn define the characterization of coral species living in their specific environments. For instance, ocean warming and acidification are causing drastic changes that affect the sustainability of coral reef ecosystems, including food availability and services provided for humans [15, 39].

In this chapter, the nutrition in corals including recent advancements in the definition of coral health, energy budget, and performance under current environmental challenges of climate changes is explained, and the implications on the survival of corals are highlighted with the aim to define future reef habitats as refugia.

## 2. Coral nutrition

Corals are unique organisms capable of taking in nutrients and gaining energy for their metabolic processes, acting like nearly every trophic level in the marine ecosystem. For instance, it has been demonstrated that corals can behave simultaneously as: *i)* primary producer, by fixing carbon and producing biomass through photosynthesis; *ii)* primary consumer, by utilizing the products of photosynthesis; and *iii)* secondary and tertiary consumers, by degrading the substrate or taking in dissolved organic matter through the ingestion of zooplankton and bacteria [40]. Therefore, corals can optimize the feeding modes to contribute to the total daily energy budget according to the surrounding conditions. However, these processes depend mostly on light and food availability, and this determines the trophic niches and the metabolic plasticity to environmental changes [27].

### 2.1 Trophic niches

The diet of corals, however, goes beyond a fixed trophic strategy based on the morphology of polyps and corallites [41]. There is a need to consider trophic plasticity as a critical factor of resistance to environmental stress [42]. The position of corals within the reef food web could be considered as the "*movement or storage of energy or materials*" [43] to identify the ecological functions of corals within such ecosystems. For example, energy allocation can shift from growth and reproduction under optimal environmental conditions to prioritizing long-term survival by depleting energy reserves under stress and shifting to anaerobic respiration of the coral hosts [44]. This concept has been applied to aquatic invertebrates, including corals, and the investigation of energy reserves based on the trophic strategy of corals is important to understand their metabolic responses to climate change, with significant implications on future coral refugia. Recently, scientific techniques have been advanced to define the trophic position of corals, and their plasticity within the reef niches. In this sense, the analysis of stable isotopes of carbon ($\delta^{13}C$) and nitrogen ($\delta^{15}N$) in coral samples allows the identification of either heterotrophic, autotrophic, or mixotrophic corals based on nutritional fluxes between coral hosts and endosymbionts [30, 45]. In this way, it is possible to recognize changes in trophic strategy (i.e. trophic plasticity) among different coral species living under the same environmental conditions [41]. This is particularly important in the ecological success of corals living in subtropical areas (**Figure 2**) [23] due to their capacity of using different nutrient sources to gain metabolic energy. Although the heterotrophic strategy has been suggested as a trophic key to enhancing bleaching resistance [45], the identification of the Symbiodiniaceae endosymbiont species and their role in acquiring essential nutrients needs to be considered as a thermotolerance feature of future corals [34].

### 2.2 Coral feeding

A substantial amount of energy in scleractinian corals is acquired through heterotrophy, which has become a key process to determine the resistance of corals to adverse conditions. Through heterotrophy, more energy for metabolic needs is available and therefore enhances the capability to resist stress events, which promotes bleaching resilience, raises protein levels, and in turn, supports the endosymbionts' physiological status [46]. Trophic differences are recognizable in the feeding rates of different species of corals. *Pocillopora* spp., for instance, have a higher capability to capture *Artemia* nauplii than *Acropora* spp. with different morphology, polyp extension, and feeding capacity. This, in turn, increases the growth rates and photosynthetic efficiency of endosymbionts, enhances the resistance to bleaching, and improves the general health status of corals [46, 47]. The key role of feeding is therefore not only related to increasing energy of the hosts but also supporting the processes involved in the endosymbionts, including photosynthesis and remineralization of organic matter. The transfer of nutrients between the hosts and endosymbionts has been recognized as inclusive of the mutualistic symbiosis of corals [48, 49]. The active intake of external organic matter, indeed, drives the acquisition of nitrogen fundamental for supporting the symbiont diversity and chlorophyll concentrations, conditions that are favorable to boost tissue growth, productivity, and calcification rates [50, 51]. Heterotrophy, however, is dependent on light, turbidity, and temperature, and can contribute to up to 35% of the metabolic energy in healthy corals [49] through assimilation of essential organic compounds of energy reserves such as lipids, proteins, and carbohydrates that cannot be acquired by photosynthesis only [46]. These are important biomarkers of coral physiology under climatic stress.

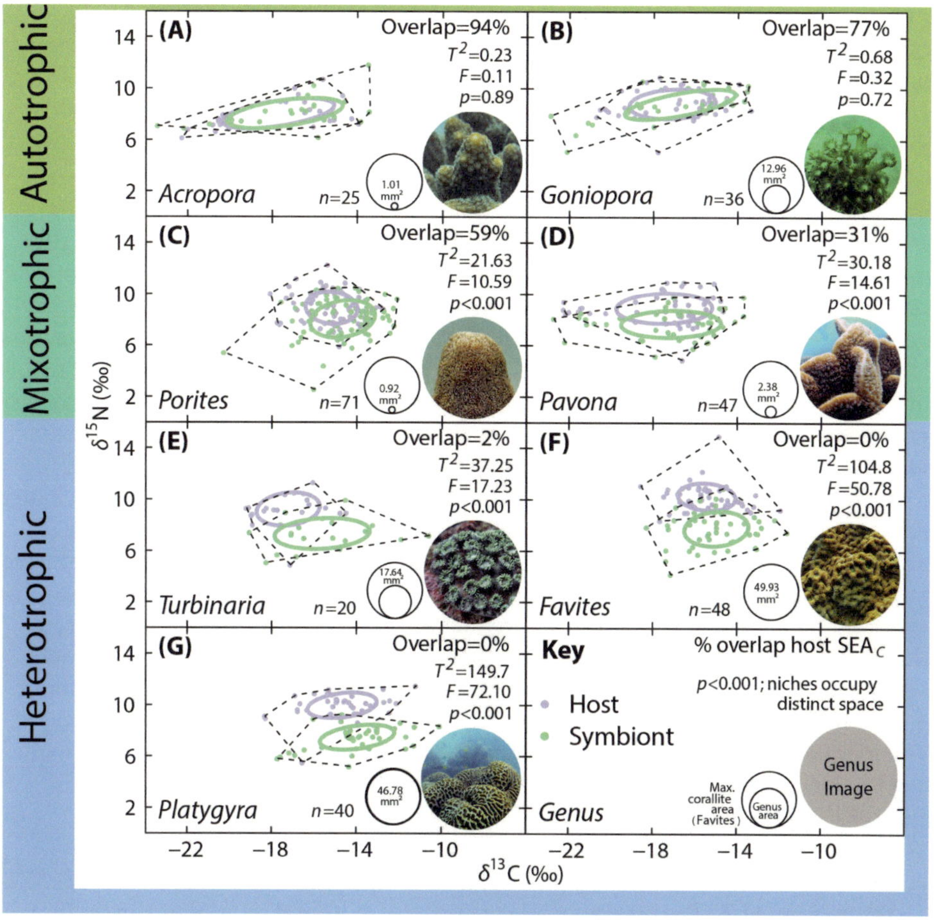


**Figure 2.**
*Trophic niches of the coral hosts (purple) and photosynthetic endosymbionts (green) analyzed with the stable isotope Bayesian analysis (SIBER). The overlapping of $\delta^{13}C$ and $\delta^{15}N$ indicates different trophic strategies between autotrophy, mixotrophy, and heterotrophy. From Conti-Jerpe et al. [30]. Reprinted with permission from AAAS.*

### 2.3 Energy reserves

**Lipids** are a fundamental component for the metabolic needs of corals and account for at least 30% of the energy reserves in corals [52]. These molecules indeed constitute much of the coral body composition, cells, and subcellular organelles [53]. The composition of fatty acids is mostly species-specific, and they are used as a chemotaxonomic indicator of the metabolic status of corals, and to trace the nutritional input of corals [54]. Polyunsaturated fatty acids (PUFAs) are one of the major lipids, and they are widely used as an indicator of dietary sources in heterotrophic corals for coping with metabolic stress [55]. It has been reported that when corals are exposed to high irradiance or heat stress, the intracellular PUFA content could decrease by up to 75% [56]. PUFAs are transported into coral compartments through feeding of zooplankton [57], and they reach the tissue after 1–2 weeks of incorporation [47]. The photosynthetic product supplied by endosymbionts is a second source of PUFAs [46]. Moreover, the PUFA content can vary depending on the depth, season, and niche distribution of zooplankton in reef ecosystems [58] and in temperate waters [59].

**Proteins** are another key component of the coral cellular physiology, since they are involved in enzymatic catalysis, cellular transportation, immunity, and growth. Heterotrophic corals are able to have higher growth rates with higher tissue protein and lipid contents, which in turn facilitate calcification, tissue synthesis, and the formation of more polyps. Interestingly, recent studies have shown that scleractinian corals are not uniform in their morphology, and different parts of corals (core, branches, etc.) in different regions are functionally specialized to meet the specific energetic demands from coral surfaces to branches [60]. The field of proteomics, i.e. identification and quantification of cellular proteins, has been recently advancing with the aim to identify key physiological processes for uncovering cellular responses under environmental changes [61]. The expression of cellular stress molecular biomarkers represents a useful diagnostic tool to analyze changes in the cellular structural integrity and in the functional cellular pathways [62]. For instance, changes in the expression of heat-shock proteins (Hsps) are emerging as ubiquitous and putative markers of stress in corals [63–67]. Hsps are molecular chaperones that have vital cellular homeostatic and cytoprotective functions and represent one of the most important defense mechanisms of all organisms [68]. Hsps are present in different cellular compartments where they participate in various housekeeping tasks, such as proper protein folding, translocation of proteins between cellular compartments, and assembly of protein complexes [69, 70]. Hsps are classified by molecular weight in major chaperone families (Hsp40, Hsp60, Hsp70, Hsp90, Hsp100, and the small Hsps), which include several members with specific intracellular localization and functions [71].

**Carbohydrates** form an important component of the coral energy reserves because they are involved in the production of energetic metabolites such as ATP [72]. The production of ATP is crucial in all functions of the coral physiology, including cellular productivity, functioning, growth, and reproduction [44]. Carbohydrates in corals are acquired by both active capturing of food through heterotrophy and by translocation from the photosynthetic process. Excess carbohydrates can be released externally as a mucoid matrix [73, 74] or stored in tissue as lipids [75]. The coral mucus, a mucoid polysaccharide external layer, plays multiple roles. It serves as an attracting layer to capture food, a food source for bacteria by trapping organic particles [76, 77], and it also creates a probiotic pool to protect the holobiont from external pathogens and viruses [37, 78, 79].

For these reasons, carbohydrates are also considered an indicator of the coral health status. Indeed, the levels of intracellular carbohydrates indicate the capacity of corals to modulate thermal stress, and therefore indicate the thermotolerance of corals [35]. These findings suggest that elevated levels of carbohydrates are related to higher adaptation to future climatic conditions and reduced bleaching susceptibility to extreme events [80, 81].

## 3. Responses of corals to environmental changes

The coral holobiont is capable of modulating its metabolic processes to dissipate or gain energy from different sources depending on nutrient availability. However, corals need to adopt special measures to face climatic change which is modifying the physicochemical and nutrient environment.

Anthropogenic activities are increasing levels of carbon dioxide ($CO_2$) in the atmosphere, leading to global warming and more frequent heatwaves, which are apparently associated with reduced rates of growth, calcification, and other functional traits, such as skeletal density, volume, and size [82–84]. These changes may in turn induce coral bleaching and mass mortality, and in the longer term, decline in coral biodiversity [81]. About a quarter of the atmospheric $CO_2$ dissolves in the ocean and reduces the seawater pH and carbonate saturation state, a process which is commonly known as ocean acidification. Ocean surface pH is expected to decrease by 0.3 units by 2100 under the RCP8.5 scenario [85, 86]. This, accompanied by cellular oxidative stress, can reduce the capacity of scleractinian corals and other calcifying organisms to build their calcium carbonate skeletons [87]. Besides global changes, human activities are responsible for multiple local pressures on marine ecosystems, specifically on corals. Coastal water quality declines in overpopulated areas, where high levels of dissolved inorganic nutrients cause eutrophication, sedimentation, and turbidity events [88–90]. The alteration of water conditions in the surface layer results in changes in the nutrient equilibrium (e.g. in the Redfield stoichiometry of C:N:P elements), which have brought about imbalanced physiological status of corals and their symbionts, and consequently increased frequency and severity of mass coral bleaching events [91, 92].

### 3.1 Responses to thermal stress

Among the plethora of stress factors, the rising of sea surface temperature due to global warming is certainly recognized as the prominent cause of coral bleaching inducing mass coral mortality [93, 94]. However, variable spatial and temporal patterns of mass bleaching have been extensively observed and can be generated by several factors that, by operating in combination, can determine different sensitivities of coral taxa to stressors [93, 95, 96]. For example, the extent of bleaching can depend on the duration and frequency of thermal anomalies and on-site-specific environmental conditions [97–100]. Nevertheless, several studies have pointed out that intrinsic factors of corals, including their morphological and physiological characteristics, play a fundamental role in determining the different levels of physiological resistance to environmental stress. In this context, the identity and clade of the Symbiodiniaceae partner may affect the coral susceptibility to unfavorable conditions [101–102].

Attempts to understand the differences in the response of corals to stress have also focused on coral physical properties, such as the coral morphology and tissue thickness, which influence growth, metabolic rates, and metabolite exchange across boundary layers and host $CO_2$ supply strategies [103, 104]. Therefore, faster growing branching taxa with thinner tissues appear more susceptible to elevated temperature than slower growing massive taxa with thicker tissues due to the latter's lower photoprotective capacity and ability to remove oxygen radicals generated during metabolic stress [103, 105].

### 3.2 Cellular stress responses

The cellular stress responses of corals are involved in driving spatial and temporal patterns of coral bleaching at both intra- and inter-specific levels. As sessile organisms, corals cannot easily migrate to new environmental optima. Therefore, in order to cope with perturbations, they rely mainly on the efficiency of their molecular and cellular mechanisms, which represent the first line of defense in reducing the harmful effects of unfavorable conditions [106–108]. The capacity of acquiring metabolic energy from autotrophy rather than heterotrophy, and vice versa, is the key to a successful symbiotic relationship in corals. However, decreased capacity to take in nutrient has been observed during thermal stress along with reduced levels of dissolved inorganic nutrients [109], impairing the assimilation of carbon and nitrogen from the hosts' heterotrophy, and inducing starvation and parasitism [38]. Recent studies have identified positive correlation between the trophic status of host and endosymbionts in *Stylophora* proving that increased photosynthetic performance is related to the amount of inorganic nutrients assimilated and translocated between hosts and symbionts, and indicating that functional heterotrophy requires essential nutrients acquired through photosynthesis [42]. On the other hand, when facing an elevated nutrient concentration, corals might exhibit thermotolerance by maintaining symbionts as an autotrophic nutrient supply for the entire holobiont [42, 109] which suggests that nitrogen enrichment might enhance the resilience of corals to thermal stress [108]. In contrast, excessive nutrients in seawater have been linked to anthropogenic activities along the coastline, such as sewage plants, dredging, and agricultural activities. These conditions are typical of rapid urbanization and industrialization and are becoming critical for biogenic habitats near urban areas [7, 23, 110, 111]. Therefore, it is critical to identify and understand the trophic plasticity of corals in relation to nutrient availability and environmental stressors.

At the cellular level, Hsps are expressed under normal physiological conditions for maintenance of normal protein folding, signal transduction, and/or normal development [112]. Moreover, their expression is upregulated as a consequence of exposure to conditions that perturb cellular protein structures [69]. The expression of Hsps, and in particular that of Hsp70 and Hsp60, has been extensively analyzed in corals subjected to extreme temperatures and bleaching conditions [113–118]. However, Hsp modulation has also been observed in corals exposed to elevated light intensity [119, 120], salinity change [121, 122], and xenobiotics/nutrient enrichments [62, 118, 123]. Recently, it has been observed that Hsps may also play a role in the immune system of corals in response to pathogen invasion [65, 124]. In most of these studies, higher Hsp levels in corals generally infer higher protection toward environmental stressors and bleaching. For instance, corals with different susceptibilities to bleaching differ in their Hsp expression levels, with the bleaching-tolerant corals exhibiting higher expression levels than the bleaching-susceptible ones [96, 119]. A recent field study showed that healthy coral colonies of *Goniopora lobata* Milne Edwards [125] and *Porites lobata* Dana [126] of the central Red Sea had higher Hsp70 and Hsp60 levels than their respective naturally bleached counterparts [66].

In addition, high Hsp levels also contribute to corals adaptation to extreme conditions, such as those characterizing the shallow lagoons of the Maldivian reefs. There, despite the remarkable daily fluctuations in temperature and light and the regular exposure to higher temperature/light regimes than surrounding waters, which can exceed their tolerance threshold and would ordinarily induce stress and bleaching, Hsp modulation seems to play a protective role to prevent the rupture of symbiosis of corals [120]. Likewise, the Hsp levels have been found to be significantly higher in bleaching-tolerant corals originating from highly variable environments compared to corals that live in more stable environments. On Ofu Island (American Samoa), colonies of *Acropora hyacinthus* Dana [126] from adjacent tidal pools with high daily thermal fluctuations were found to be more thermotolerant and had constitutively higher levels of Hsp70 gene compared to bleaching-sensitive colonies from less thermally variable pools [127]. Similarly, corals from inshore reefs of Florida bay *Porites astreoides,* Lamarck [128] were subjected to temperature fluctuations and appeared to have higher levels of Hsp genes than the offshore corals [107]. During the bleaching event of 2016, the near-shore colonies in Mauritius did not bleach and had significantly higher relative levels of both Hsp70 and 60 genes and protein compared to bleached reef colonies, indicating that the modulation of these Hsps was involved in local acclimatization of corals to their environments [96]. However, it is important to consider that prior exposure to sublethal environment stress (preconditioning) that resulted in later tolerance to bleaching temperatures [129] and changes in the expression of specific genes, such as those of Hsps, have been associated with this thermal tolerance plasticity [130, 131]. For example, the preconditioning of *Acropora millepora* (Ehrenberg, 1834) colonies to heat stress accounted for increased gene expression and tolerance to bleaching compared to nonpreconditioned colonies [63]. Overall, the expression pattern of Hsps and the amplitude of their modulation may show species-specific characteristics, which may reflect different mechanisms and abilities of stress response.

### 3.3 Thermal performance

Metabolic performance, in particular the thermal performance curve (TPC) which defines the nonlinear relationship of organismal metabolism versus a given source of stress, is another parameter to consider when coral health is concerned. When the metabolic response of corals to low/high temperature is considered, the TPC can be applied to quantify the response of a coral species to thermal stress [132]. Moreover, through the TPC, it is possible to measure the maximum level of such performance, the optimal conditions of temperature, and the capacity of resistance to temperature variation (e.g. thermal breadth). The shape of such curve and its relative breadth will determine the metabolic plasticity of organisms (corals) to temperature variations. This can be used to define the physiological performance of corals and compare their specific responses in the subtropical area to indicate physiological adaptation of corals to living conditions and the challenges that subtropical corals face when optimizing their productivity in subtropical environments [133]. The heritability of coral traits must also be considered in the framework of coral adaptation to future conditions of climate change to better predict the evolution of corals in suboptimal conditions [134].

## 4. Future coral refugia

Coral reefs are often described as biogenic structures which provide nutriments and services to the marine ecosystems formed in oligotrophic areas (i.e. low dissolved nutrients and clear water) with stable environmental conditions. These features are usually optimal for bioconstruction, such as corals, to capture carbonates from seawater and sustain the metabolic energy needed for growth and reproduction [93, 135]. The capacity of corals to modulate their metabolism according to surrounding conditions is the key for their success. Scleractinian corals are thriving also in the so-called marginal reefs, where thermal and salinity anomalies, eutrophication, and elevated sedimentation rates are the causes of metabolic expenditures and, eventually, stress [136–139].

Marginal reefs are located at high latitudes of subtropical areas and near megalopolis. Corals living in these areas receive multiple pressures from local stressors together with global changes, although the processes involved in these ecosystems operate at different spatial (i.e. geographical) and temporal (i.e. frequency of stress events) scales compared to tropical reefs. In this context, it is important to consider how natural evolution, affected by human pressures, has shaped the coral species living in these areas, and how the marginal reefs can act as refuge area for future conditions. Refugia are considered as those areas with the ability to provide protection from multiple stressors [140], and in this case coral refugia are identified as areas where long-term stressors are low that less likely to influence coral survival. For examples, considering the evolutionary timescales, the current marginal reefs are already serving as refugia due to their environmental conditions [9], although with reduced speciation, growth, and reproduction rates [141]. Moreover, most of the research works have focused on the short-term relief to environmental stressors, and there is a need to understand how the marginal reefs can act as refugia under the climatic scenarios of more frequent heatwave events and continuous development of coastal areas [142]. The understanding of the responses of corals in the adaptation and evolution in these areas is therefore a priority for devising conservation and restoration measures for the future coral reefs [3]. Recent studies have identified areas as future refugia from thermal stress. Corals living in environments with naturally high temperature fluctuation may have developed higher thermal tolerance to heat stress, and therefore these areas can be considered as refugia for future conditions. To represent refugia areas with a high potential to maintain the future coral biodiversity and ecosystem functions, the frequency of thermal stress events (e.g. 12-week sum of 1°C higher than the maximum monthly mean) should be less than one every 10 years [143]. Future warming conditions and more heatwaves might result in too frequent thermal stress events and leave no room for those corals and other marine organisms that live in thermal refugia to adapt. The biological responses to the chronic development of ocean warming will be critical to determine the effectiveness of high-latitude reefs as the thermal refugia [143].

## 5. Conclusion

Coral reefs have very high biodiversity values and provide important ecosystem services, with the capacity to resist anthropogenic stress by modulating their energetic budgets as described in this chapter. Current major threats to them are caused by increasing seawater temperature (ocean warming) and reduced pH level (ocean acidification), which cause reduction in survival, calcification, growth, and photosynthesis in several marine taxa [138, 144] with the levels of impacts depending on morphology and the feeding capacity of corals [45]. There are global consequences of this reduced capacity of reef ecosystems to provide crucial services, such as reduced fishing capacity and unsustainable management of marine reserves [145, 146]. A deep understanding of the multiple interactions between stressors and mitigators will be crucial to define the trophic plasticity and reef responses under the future environmental changes.

## Acknowledgements

This study was partially funded by the Research Institute for Future Food, The Hong Kong Polytechnic University.

## Conflict of interest

The authors declare no conflict of interest.

## Author details

Walter Dellisanti[1], Davide Seveso[2,3] and James Kar-Hei Fang[1,4*]

1 Research Institute for Future Food, and Department of Applied Biology and Chemical Technology, The Hong Kong Polytechnic University, Hong Kong SAR, China

2 Department of Earth and Environmental Sciences, University of Milano, Bicocca, Italy

3 Marine Research and High Education Center, Magoodhoo, Faafu, Maldives

4 State Key Laboratory of Marine Pollution, City University of Hong Kong, Hong Kong SAR, China

*Address all correspondence to: james.fang@polyu.edu.hk

## References

[1] Buhl-Mortensen L, Vanreusel A, Gooday AJ, et al. Biological structures as a source of habitat heterogeneity and biodiversity on the deep ocean margins. Marine Ecology. 2010;**31**(1):21-50. DOI: 10.1111/J.1439-0485.2010.00359.X

[2] Enochs IC, Manzello DP, Donham EM, et al. Shift from coral to macroalgae dominance on a volcanically acidified reef. Nature Climate Change. 2015;**5**(12):1083-1088. DOI: 10.1038/nclimate2758

[3] Camp EF, Schoepf V, Mumby PJ, et al. The future of coral reefs subject to rapid climate change: Lessons from natural extreme environments. Frontiers in Marine Science. 2018;**5**:1-21. DOI: 10.3389/fmars.2018.00004

[4] López-Victoria M, Zea S, Weil E. Competition for space between encrusting excavating Caribbean sponges and other coral reef organisms. Marine Ecology Progress Series. 2006;**312**:113-121. DOI: 10.3354/MEPS312113

[5] Chui APY, Yeung CW, Tsang RHL, et al. Lowered temperature and reduced salinity retarded development of early life history stages of Acropora valida from the marginal environment. Regional Studies in Marine Science. 2016;**8**:430-438. DOI: 10.1016/j.rsma.2016.04.004

[6] Kurihara H, Wouters J, Yasuda N. Seasonal calcification of the coral Acropora digitifera from a subtropical marginal Okinawa reef under ocean acidification. Coral Reefs. 2019;**38**(3):443-454. DOI: 10.1007/s00338-019-01794-9

[7] Heery EC, Hoeksema BW, Browne NK, et al. Urban coral reefs: Degradation and resilience of hard coral assemblages in coastal cities of east and Southeast Asia. Marine Pollution Bulletin. 2018;**135**:654-681. DOI: 10.1016/j.marpolbul.2018.07.041

[8] Chimienti G, Mastrototaro F, D'Onghia G. Mesophotic and deep-sea vulnerable coral habitats of the Mediterranean Sea: Overview and conservation perspectives. In: Advances in the Studies of the Benthic Zone. Rijeka: IntechOpen; 2019. DOI: 10.5772/intechopen.90024

[9] Lybolt M, Neil D, Zhao J, Feng Y, Yu K-F, Pandolfi J. Instability in a marginal coral reef: The shift from natural variability to a human-dominated seascape. Frontiers in Ecology and the Environment. 2011;**9**(3):154-160. DOI: 10.1890/090176

[10] Pratchett MS, Anderson KD, Hoogenboom MO, Widman E, Baird AH, Pandolfi JM, et al. Spatial, temporal and taxonomic variation in coral growth - implications for the structure and function of coral reef ecosystems. Oceanography and Marine Biology: An Annual Review. 2015;**53**:215-295

[11] Roberts CM, McClean CJ, Veron JEN, et al. Marine biodiversity hotspots and conservation priorities for tropical reefs. Science. 2002;**295**(5558):1280-1284. DOI: 10.1126/science.1067728

[12] Knowlton N, Brainard RE, Fisher R, Moews M, Plaisance L, Caley MJ. Coral reef biodiversity. In: McIntyre AD, editor. Life in the World's Oceans. Oxford, UK: Wiley-Blackwell; 2010. pp. 65-78. DOI: 10.1002/9781444325508.ch4

[13] Van Zanten BT, Van Beukering PJH, Wagtendonk AJ. Coastal protection

by coral reefs: A framework for spatial assessment and economic valuation. Ocean and Coastal Management. 2014;**96**:94-103. DOI: 10.1016/j.ocecoaman.2014.05.001

[14] Hoegh-Guldberg O, Mumby PJ, Hooten AJ, et al. Coral reefs under rapid climate change and ocean acidification. Science. 2007;**318**(5857):1737-1743. DOI: 10.1126/science.1152509

[15] Hughes TP, Barnes ML, Bellwood DR, et al. Coral reefs in the Anthropocene. Nature. 2017;**546**(7656):82-90. DOI: 10.1038/nature22901

[16] Moss RH, Edmonds JA, Hibbard KA, et al. The next generation of scenarios for climate change research and assessment. Nature. 2010;**463**(7282):747-756. DOI: 10.1038/nature08823

[17] Ellis JI, Jamil T, Anlauf H, et al. Multiple stressor effects on coral reef ecosystems. Global Change Biology. 2019;**25**(12):4131-4146. DOI: 10.1111/GCB.14819

[18] Dellisanti W, Chung JTH, Chow CFY, Wu J, Wells ML, Chan LL. Experimental techniques to assess coral physiology in situ under global and local stressors: Current approaches and novel insights. Frontiers in Physiology. 2021;**12**:656562. DOI: 10.3389/fphys.2021.656562

[19] Nyström M, Folke C, Moberg F. Coral reef disturbance and resilience in a human-dominated environment. Trends in Ecology & Evolution. 2000;**15**(10):413-417. DOI: 10.1016/S0169-5347(00)01948-0

[20] Szmant AM. Nutrient enrichment on coral reefs: Is it a major cause of coral reef decline? Estuaries. 2002;**25**(4):743-766. DOI: 10.1007/BF02804903

[21] Fabricius KE, De'ath, G. Identifying ecological change and its cause: A case study on coral reefs. Ecological Applications. 2004;**14**(5):1448-1465. DOI: 10.1890/03-5320

[22] Bruno JF, Selig ER, Casey KS, et al. Thermal stress and coral cover as drivers of coral disease outbreaks. PLoS Biology. 2007;**5**(6):e124. DOI: 10.1371/JOURNAL.PBIO.0050124

[23] Lachs L, Johari NAM, Le DQ, et al. Effects of tourism-derived sewage on coral reefs: Isotopic assessments identify effective bioindicators. Marine Pollution Bulletin. 2019;**148**:85-96. DOI: 10.1016/J.MARPOLBUL.2019.07.059

[24] Metian M, Hédouin L, Ferrier-Pagès C, et al. Metal bioconcentration in the scleractinian coral Stylophora pistillata: Investigating the role of different components of the holobiont using radiotracers. Environmental Monitoring and Assessment. 2015;**187**(4):1-10. DOI: 10.1007/s10661-015-4383-z

[25] LaJeunesse TC, Parkinson JE, Gabrielson PW, et al. Systematic revision of Symbiodiniaceae highlights the antiquity and diversity of coral endosymbionts. Current Biology. 2018;**28**(16):2570-2580. DOI: 10.1016/j.cub.2018.07.008

[26] Fox MD, Elliott Smith EA, Smith JE, Newsome SD. Trophic plasticity in a common reef-building coral: Insights from δ13C analysis of essential amino acids. Functional Ecology. 2019;**33**(11):2203-2214. DOI: 10.1111/1365-2435.13441

[27] Grottoli AG, Tchernov D, Winters G. Physiological and biogeochemical responses of super-corals to thermal stress from the northern Gulf of Aqaba, Red Sea. Frontiers in Marine Science. 2017;**4**:215. DOI: 10.3389/fmars.2017.00215

[28] Wangpraseurt D, Pernice M, Guagliardo P, et al. Light microenvironment and single-cell gradients of carbon fixation in tissues of symbiont-bearing corals. The ISME Journal. 2015;**10**(3):788-792. DOI: 10.1038/ismej.2015.133

[29] Carbonne C, Teixidó N, Moore B, et al. Two temperate corals are tolerant to low pH regardless of previous exposure to natural $CO_2$ vents. Limnology and Oceanography. 2021;**66**(11):4046-4061. DOI: 10.1002/LNO.11942

[30] Conti-Jerpe IE, Thompson PD, Wong CWM, et al. Trophic strategy and bleaching resistance in reef-building corals. Science Advances. 2020;**6**(15):5443-5453. DOI: 10.1126/sciadv.aaz5443

[31] Gardella DJ, Edmunds PJ. The oxygen microenvironment adjacent to the tissue of the scleractinian Dichocoenia stokesii and its effects on symbiont metabolism. Marine Biology. 1999;**135**(2):289-295. DOI: 10.1007/s002270050626

[32] Linsmayer LB, Deheyn DD, Tomanek L, Tresguerres M. Dynamic regulation of coral energy metabolism throughout the diel cycle. Scientific Reports. 2020;**10**(1):1-11. DOI: 10.1038/s41598-020-76828-2

[33] Comeau S, Tambutté E, Carpenter RC, et al. Coral calcifying fluid pH is modulated by seawater carbonate chemistry not solely seawater pH. Proceedings of the Royal Society B: Biological Sciences. 1847;**2017**:284. DOI: 10.1098/rspb.2016.1669

[34] Reich HG, Tu WC, Rodriguez IB, et al. Iron availability modulates the response of Endosymbiotic dinoflagellates to heat stress. Journal of Phycology. 2021;**57**(1):3-13. DOI: 10.1111/JPY.13078

[35] Kochman NR, Grover R, Rottier C, Ferrier-Pages C, Fine M. The reef building coral Stylophora pistillata uses stored carbohydrates to maintain ATP levels under thermal stress. Coral Reefs. 2021;**40**(5):1473-1485. DOI: 10.1007/S00338-021-02174-Y/FIGURES/4

[36] Thurber RLV, Correa AMS. Viruses of reef-building scleractinian corals. Journal of Experimental Marine Biology and Ecology. 2011;**408**(1-2):102-113. DOI: 10.1016/j.jembe.2011.07.030

[37] Peixoto RS, Rosado PM, de Assis Leite DC, Rosado AS, Bourne DG. Beneficial microorganisms for corals (BMC): Proposed mechanisms for coral health and resilience. Frontiers in Microbiology. 2017;**8**:1-16. DOI: 10.3389/fmicb.2017.00341

[38] Baker DM, Freeman CJ, Wong JCYY, Fogel ML, Knowlton N. Climate change promotes parasitism in a coral symbiosis. The ISME Journal. 2018;**12**(3):921-930. DOI: 10.1038/s41396-018-0046-8

[39] Putnam HM, Barott KL, Ainsworth TD, Gates RD. The vulnerability and resilience of reef-building corals. Current Biology. 2017;**27**(11):R528-R540. DOI: 10.1016/j.cub.2017.04.047

[40] Muscatine L, Porter JW. Reef corals: Mutualistic symbioses adapted to nutrient-poor environments. Bioscience. 1977;**27**(7):454-460. DOI: 10.2307/1297526

[41] Sturaro N, Hsieh YE, Chen Q, Wang PL, Denis V. Trophic plasticity of mixotrophic corals under contrasting environments. Functional Ecology. 2021;**35**(12):2841-2855. DOI: 10.1111/1365-2435.13924/SUPPINFO

[42] Ezzat L, Maguer JF, Grover R, Rottier C, Tremblay P, Ferrier-Pagès C.

Nutrient starvation impairs the trophic plasticity of reef-building corals under ocean warming. Functional Ecology. 2019;**33**(4):643-653. DOI: 10.1111/1365-2435.13285/SUPPINFO

[43] Bellwood DR, Streit RP, Brandl SJ, Tebbett SB. The meaning of the term 'function' in ecology: A coral reef perspective. Functional Ecology. 2019;**33**(6):948-961. DOI: 10.1111/1365-2435.13265/SUPPINFO

[44] Sokolova IM, Frederich M, Bagwe R, Lannig G, Sukhotin AA. Energy homeostasis as an integrative tool for assessing limits of environmental stress tolerance in aquatic invertebrates. Marine Environmental Research. 2012;**79**:1-15. DOI: 10.1016/J.MARENVRES.2012.04.003

[45] Ferrier-Pagès C, Sauzéat L, Balter V. Coral bleaching is linked to the capacity of the animal host to supply essential metals to the symbionts. Global Change Biology. 2018;**24**(7):3145-3157. DOI: 10.1111/GCB.14141

[46] Huang YL, Mayfield AB, Fan TY. Effects of feeding on the physiological performance of the stony coral Pocillopora acuta. Scientific Reports. 2020;**10**(1):1-12. DOI: 10.1038/s41598-020-76451-1

[47] Lim CS, Bachok Z, Hii YS. Effects of supplementary polyunsaturated fatty acids on the health of the scleractinian coral Galaxea fascicularis (Linnaeus, 1767). Journal of Experimental Marine Biology and Ecology. 2017;**491**:1-8. DOI: 10.1016/j.jembe.2017.02.009

[48] Anthony KRN, Fabricius KE. Shifting roles of heterotrophy and autotrophy in coral energetics under varying turbidity. Journal of Experimental Marine Biology and Ecology. 2000;**252**(2):221-253. DOI: 10.1016/S0022-0981(00)00237-9

[49] Houlbrèque F, Ferrier-Pagès C. Heterotrophy in tropical Scleractinian corals. Biological Reviews. 2009;**84**(1):1-17. DOI: 10.1111/J.1469-185X.2008.00058.X

[50] Ferrier-Pagès C, Witting J, Tambutté E, Sebens KP. Effect of natural zooplankton feeding on the tissue and skeletal growth of the scleractinian coral Stylophora pistillata. Coral Reefs. 2003;**22**(3):229-240. DOI: 10.1007/S00338-003-0312-7/TABLES/3

[51] Yu X, Jiang L, Gan J, et al. Effects of feeding on production, body composition and fatty acid profile of scleractinian coral Galaxea fascicularis. Aquaculture Reports. 2021;**21**:100871. DOI: 10.1016/J.AQREP.2021.100871

[52] Parrish CC. Lipids in marine ecosystems. ISRN Oceanography. 2013;**2013**:1-16. DOI: 10.5402/2013/604045

[53] Liu CY, Zhang F, Sun YF, Yu XL, Huang H. Effects of nitrate enrichment on respiration, photosynthesis, and fatty acid composition of reef coral *Pocillopora damicornis* larvae. Frontiers in Marine Science. 2020;**7**:531. DOI: 10.3389/FMARS.2020.00531/BIBTEX

[54] Mies M, Güth AZ, Tenório AA, et al. In situ shifts of predominance between autotrophic and heterotrophic feeding in the reef-building coral Mussismilia hispida: An approach using fatty acid trophic markers. Coral Reefs. 2018;**37**(3):677-689. DOI: 10.1007/S00338-018-1692-Z/FIGURES/7

[55] Seemann J, Sawall Y, Auel H, Richter C. The use of lipids and fatty acids to measure the trophic plasticity of the coral Stylophora subseriata. Lipids.

2013;**48**(3):275-286. DOI: 10.1007/S11745-012-3747-1

[56] Bachok Z, Mfilinge P, Tsuchiya M. Characterization of fatty acid composition in healthy and bleached corals from Okinawa, Japan. Coral Reefs. 2006;**25**(4):545-554. DOI: 10.1007/S00338-006-0130-9/TABLES/3

[57] Tagliafico A, Rudd D, Rangel MS, et al. Lipid-enriched diets reduce the impacts of thermal stress in corals. Marine Ecology Progress Series. 2017;**573**:129-141. DOI: 10.3354/MEPS12177

[58] Latyshev NA, Naumenko NV, Svetashev VI, Latypov YY. Fatty acids of reef-building corals. Marine Ecology Progress Series. 1991;**76**:295-301

[59] Rodolfo-Metalpa R, Peirano A, Houlbrèque F, Abbate M, Ferrier-Pagès C. Effects of temperature, light and heterotrophy on the growth rate and budding of the temperate coral Cladocora caespitosa. Coral Reefs. 2008;**27**(1):17-25. DOI: 10.1007/S00338-007-0283-1/FIGURES/5

[60] Conlan JA, Humphrey CA, Severati A, Francis DS. Intra-colonial diversity in the scleractinian coral, *Acropora millepora* : Identifying the nutritional gradients underlying physiological integration and compartmentalised functioning. PeerJ. 2018;**6**:e4239. DOI: 10.7717/peerj.4239

[61] Ma H, Liao H, Dellisanti W, Sun Y, Chan LL, Zhang L. Characterizing the host coral proteome of Platygyra carnosa using suspension trapping (S-trap). Journal of Proteome Research. 2021;**20**(3):1783-1791. DOI: 10.1021/ACS.JPROTEOME.0C00812

[62] Downs CA. Cellular diagnostics and its application to aquatic and marine toxicology. In: Ostrander GK, editor. Techniques in Aquatic Toxicology. Vol. 2. Boca Raton, USA: CRC Press; 2010

[63] Bellantuono AJ, Granados-Cifuentes C, Miller DJ, Hoegh-Guldberg O, Rodriguez-Lanetty M. Coral thermal tolerance: Tuning gene expression to resist thermal stress. PLoS One. 2012;**7**(11):50685. DOI: 10.1371/JOURNAL.PONE.0050685

[64] Louis YD, Bhagooli R, Kenkel CD, Baker AC, Dyall SD. Gene expression biomarkers of heat stress in scleractinian corals: Promises and limitations. Comparative Biochemistry and Physiology, Part C: Toxicology & Pharmacology. 2017;**191**:63-77. DOI: 10.1016/J.CBPC.2016.08.007

[65] Seveso D, Montano S, Reggente MAL, et al. The cellular stress response of the scleractinian coral Goniopora columna during the progression of the black band disease. Cell Stress & Chaperones. 2017;**22**(2):225-236. DOI: 10.1007/S12192-016-0756-7/FIGURES/1

[66] Seveso D, Arrigoni R, Montano S, et al. Investigating the heat shock protein response involved in coral bleaching across scleractinian species in the Central Red Sea. Coral Reefs. 2020;**39**(1):85-98. DOI: 10.1007/S00338-019-01878-6/FIGURES/1

[67] Traylor-Knowles N, Rose NH, Sheets EA, Palumbi SR. Early transcriptional responses during heat stress in the coral Acropora hyacinthus. The Biological Bulletin. 2017;**232**(2):91-100. DOI: 10.1086/692717/ASSET/IMAGES/LARGE/FGA1.JPEG

[68] Kumsta C, Jakob U. Redox-regulated chaperones. Biochemistry. 2009;**48**(22):4666-4676. DOI: 10.1021/BI9003556

[69] Balchin D, Hayer-Hartl M, Hartl FU. In vivo aspects of protein folding and quality

control. Science. 2016;**353**(6294):4354. DOI: 10.1126/science.aac4354

[70] Rosenzweig R, Nillegoda NB, Mayer MP, Bukau B. The Hsp70 chaperone network. Nature Reviews. Molecular Cell Biology. 2019;**20**(11):665-680. DOI: 10.1038/S41580-019-0133-3

[71] Pockley AG, Georgiades A, Thulin T, De Faire U, Frostegård J. Serum heat shock protein 70 levels predict the development of atherosclerosis in subjects with established hypertension. Hypertension. 2003;**42**(3):235-238. DOI: 10.1161/01.HYP.0000086522.13672.23

[72] Hardie DG, Scott JW, Pan DA, Hudson ER. Management of cellular energy by the AMP-activated protein kinase system. FEBS Letters. 2003;**546**(1):113-120. DOI: 10.1016/S0014-5793(03)00560-X

[73] Hadaidi G, Gegner HM, Ziegler M, Voolstra CR. Carbohydrate composition of mucus from scleractinian corals from the Central Red Sea. Coral Reefs. 2019;**38**(1):21-27. DOI: 10.1007/s00338-018-01758-5

[74] Wild C, Naumann M, Niggl W, Haas A. Carbohydrate composition of mucus released by scleractinian warm- and cold-water reef corals. Aquatic Biology. 2010;**10**(1):41-45. DOI: 10.3354/AB00269

[75] Patton JS, Burris JE. Lipid synthesis and extrusion by freshly isolated zooxanthellae (symbiotic algae). Marine Biology. 1983;**75**(2-3):131-136. DOI: 10.1007/BF00405995

[76] Hadaidi G, Röthig T, Yum LK, et al. Stable mucus-associated bacterial communities in bleached and healthy corals of Porites lobata from the Arabian seas. Scientific Reports. 2017;7:1-11. DOI: 10.1038/srep45362

[77] Wild C, Woyt H, Huettel M. Influence of coral mucus on nutrient fluxes in carbonate sands. Marine Ecology Progress Series. 2005;**287**:87-98. DOI: 10.3354/meps287087

[78] Sweet M, Bythell J. The role of viruses in coral health and disease. Journal of Invertebrate Pathology. 2017;**147**:136-144. DOI: 10.1016/j.jip.2016.12.005

[79] Glasl B, Herndl GJ, Frade PR. The microbiome of coral surface mucus has a key role in mediating holobiont health and survival upon disturbance. The ISME Journal. 2016;**10**(9):2280-2292. DOI: 10.1038/ismej.2016.9

[80] Hoegh-Guldberg O, Mumby PJ, Hooten AJ, et al. Coral reefs under rapid climate change and ocean acidification. Science. 2007;**318**(5857):1737-1742. DOI: 10.1126/science.1152509

[81] Hughes TP, Kerry JT, Álvarez-Noriega M, et al. Global warming and recurrent mass bleaching of corals. Nature. 2017;**543**(7645):373-377. DOI: 10.1038/nature21707

[82] Schoepf V, D'Olivo JP, Rigal C, Jung EMU, McCulloch MT. Heat stress differentially impacts key calcification mechanisms in reef-building corals. Coral Reefs. 2021;**40**(2):459-471. DOI: 10.1007/s00338-020-02038-x

[83] Gruber N, Boyd PW, Frölicher TL, Vogt M. Biogeochemical extremes and compound events in the ocean. Nature. 2021;**600**(7889):395-407. DOI: 10.1038/s41586-021-03981-7

[84] Donovan MK, Burkepile DE, Kratochwill C, et al. Local conditions magnify coral loss after marine heatwaves. Science. 2021;**372**(6545):977-980. DOI: 10.1126/science.abd9464

[85] Ciais P, Sabine C, Bala G, Peters W. Carbon and other biogeochemical cycles. In: Stocker et al., editors. Climate Change 2013 the Physical Science Basis: Working Group I Contribution to the Fifth Assessment Report of the Intergovernmental Panel on Climate Change; United Kingdom, New York, USA: Cambridge University Press; 2013. pp. 465-570. DOI: 10.1017/CBO9781107415324.015

[86] Zeebe RE, Wolf-Gladrow D, Zakaria FZ, et al. Effects of ocean acidification on benthic processes, organisms, and ecosystems. In: Chen CA, editor. Climate Change 2013 the Physical Science Basis: Working Group I Contribution to the Fifth Assessment Report of the Intergovernmental Panel on Climate Change; United Kingdom, New York, USA: Cambridge University Press; 2013. pp. 122-153. DOI: 10.1017/CBO9781107415324.015

[87] Eyre BD, Andersson AJ, Cyronak T. Benthic coral reef calcium carbonate dissolution in an acidifying ocean. Nature Climate Change. 2014;**4**(11):969-976. DOI: 10.1038/nclimate2380

[88] Burkepile DE, Shantz AA, Adam TC, et al. Nitrogen identity drives differential impacts of nutrients on coral bleaching and mortality. Ecosystems. 2020;**23**(4):798-811. DOI: 10.1007/s10021-019-00433-2

[89] Browne NK, Precht E, Last KS, Todd PA. Photo-physiological costs associated with acute sediment stress events in three near-shore turbid water corals. Marine Ecology Progress Series. 2014;**502**(April):129-143. DOI: 10.3354/meps10714

[90] Tuttle LJ, Johnson C, Kolinski S, Minton D, Donahue MJ. How does sediment exposure affect corals? A systematic review protocol. Environmental Evidence. 2020;**9**(1):17. DOI: 10.1186/s13750-020-00200-0

[91] D'Angelo C, Wiedenmann J. Impacts of nutrient enrichment on coral reefs: New perspectives and implications for coastal management and reef survival. Current Opinion in Environment Sustainability. 2014;**7**:82-93. DOI: 10.1016/j.cosust.2013.11.029

[92] Morris LA, Voolstra CR, Quigley KM, Bourne DG, Bay LK. Nutrient availability and metabolism affect the stability of coral–Symbiodiniaceae symbioses. Trends in Microbiology. 2019;**27**(8):678-689. DOI: 10.1016/j.tim.2019.03.004

[93] Hughes TP, Anderson KD, Connolly SR, et al. Spatial and temporal patterns of mass bleaching of corals in the Anthropocene. Science. 2018;**359**(6371):80-83. DOI: 10.1126/science.aan8048

[94] Sully S, Burkepile DE, Donovan MK, Hodgson G, van Woesik R. A global analysis of coral bleaching over the past two decades. Nature Communications. 2019;**10**(1):1-5. DOI: 10.1038/s41467-019-09238-2

[95] Heron SF, Maynard JA, Van Hooidonk R, Eakin CM. Warming trends and bleaching stress of the World's coral reefs 1985-2012. Scientific Reports. 2016;**6**(1):1-14. DOI: 10.1038/srep38402

[96] Louis YD, Bhagooli R, Seveso D, et al. Local acclimatisation-driven differential gene and protein expression patterns of Hsp70 in Acropora muricata: Implications for coral tolerance to bleaching. Molecular Ecology. 2020;**29**(22):4382-4394. DOI: 10.1111/MEC.15642

[97] McClanahan TR, Maina J. Response of coral assemblages to the interaction between natural temperature variation

and rare warm-water events. Ecosystems. 2003;**6**(6):551-563. DOI: 10.1007/S10021-002-0104-X/TABLES/3

[98] Guest JR, Baird AH, Maynard JA, et al. Contrasting patterns of coral bleaching susceptibility in 2010 suggest an adaptive response to thermal stress. PLoS One. 2012;**7**(3):e33353. DOI: 10.1371/JOURNAL.PONE.0033353

[99] Howells EJ, Berkelmans R, Van Oppen MJH, Willis BL, Bay LK. Historical thermal regimes define limits to coral acclimatization. Ecology. 2013;**94**(5):1078-1088. DOI: 10.1890/12-1257.1

[100] Pratchett MS, McCowan D, Maynard JA, Heron SF. Changes in bleaching susceptibility among corals subject to ocean warming and recurrent bleaching in Moorea, French Polynesia. PLoS One. 2013;**8**(7):e70443. DOI: 10.1371/JOURNAL.PONE.0070443

[101] Hume BCC, Voolstra CR, Arif C, et al. Ancestral genetic diversity associated with the rapid spread of stress-tolerant coral symbionts in response to holocene climate change. Proceedings of the National Academy of Sciences of the United States of America. 2016;**113**(16):4416-4421. DOI: 10.1073/pnas.1601910113

[102] Ziegler M, Eguíluz VM, Duarte CM, Voolstra CR. Rare symbionts may contribute to the resilience of coral–algal assemblages. The ISME Journal. 2017;**12**(1):161-172. DOI: 10.1038/ismej.2017.151

[103] Loya Y, Sakai K, Yamazato K, Nakano Y, Sambali H, Van Woesik R. Coral bleaching: The winners and the losers. Ecology Letters. 2001;**4**(2):122-131. DOI: 10.1046/j.1461-0248.2001.00203.x

[104] Darling ES, Alvarez-Filip L, Oliver TA, Mcclanahan TR, Côté IM. Evaluating life-history strategies of reef corals from species traits. Ecology Letters. 2012;**15**(12):1378-1386. DOI: 10.1111/j.1461-0248.2012.01861.x

[105] Hoegh-Guldberg O. Climate change, coral bleaching and the future of the world's coral reefs. Marine and Freshwater Research. 1999;**50**(8):839-866. DOI: 10.1071/MF99078

[106] Brown BE. Coral bleaching: Causes and consequences. Coral Reefs. 1997;**16**(1):S129-S138. DOI: 10.1007/S003380050249

[107] Kenkel CD, Sheridan C, Leal MC, et al. Diagnostic gene expression biomarkers of coral thermal stress. Molecular Ecology Resources. 2014;**14**(4):667-678. DOI: 10.1111/1755-0998.12218

[108] Rosic N, Kaniewska P, Chan CKK, et al. Early transcriptional changes in the reef-building coral Acropora aspera in response to thermal and nutrient stress. BMC Genomics. 2014;**15**(1):1-17. DOI: 10.1186/1471-2164-15-1052/FIGURES/6

[109] Naugle MS, Oliver TA, Barshis DJ, Gates RD, Logan CA. Variation in coral Thermotolerance across a pollution gradient erodes as coral symbionts shift to more heat-tolerant genera. Frontiers in Marine Science. 2021;**8**:1670. DOI: 10.3389/FMARS.2021.760891/BIBTEX

[110] Duprey NN, Yasuhara M, Baker DM. Reefs of tomorrow: Eutrophication reduces coral biodiversity in an urbanized seascape. Global Change Biology. 2016;**22**(11):3550-3565. DOI: 10.1111/gcb.13432

[111] Duprey NN, Wang TX, Kim T, et al. Megacity development and the demise of coastal coral communities: Evidence

from coral skeleton δ $^{15}$N records in the Pearl River estuary. Global Change Biology. 2020;**26**(3):1338-1353. DOI: 10.1111/gcb.14923

[112] Chen JY, Parekh M, Seliman H, et al. Heat shock promotes inclusion body formation of mutant huntingtin (mHtt) and alleviates mHtt-induced transcription factor dysfunction. The Journal of Biological Chemistry. 2018;**293**(40):15581-15593. DOI: 10.1074/JBC.RA118.002933

[113] Brown BE, Downs CA, Dunne RP, Gibb SW. Exploring the basis of thermotolerance in the reef coral Goniastrea aspera. Marine Ecology Progress Series. 2002;**242**:119-129. DOI: 10.3354/MEPS242119

[114] Barshis DJ, Stillman JH, Gates RD, Toonen RJ, Smith LW, Birkeland C. Protein expression and genetic structure of the coral Porites lobata in an environmentally extreme Samoan back reef: Does host genotype limit phenotypic plasticity? Molecular Ecology. 2010;**19**(8):1705-1720. DOI: 10.1111/J.1365-294X.2010.04574.X

[115] Desalvo MK, Sunagawa S, Voolstra CR, Medina M. Transcriptomic responses to heat stress and bleaching in the Elkhorn coral Acropora palmata. Marine Ecology Progress Series. 2010;**402**:97-113. DOI: 10.3354/MEPS08372

[116] Kenkel CD, Traylor MR, Wiedenmann J, Salih A, Matz MV. Fluorescence of coral larvae predicts their settlement response to crustose coralline algae and reflects stress. Proceedings of the Royal Society B: Biological Sciences. 2011;**278**(1718):2691-2697. DOI: 10.1098/rspb.2010.2344

[117] Rosic NN, Pernice M, Dove S, Dunn S, Hoegh-Guldberg O. Gene expression profiles of cytosolic heat shock proteins Hsp70 and Hsp90 from symbiotic dinoflagellates in response to thermal stress: Possible implications for coral bleaching. Cell Stress & Chaperones. 2011;**16**(1):69. DOI: 10.1007/S12192-010-0222-X

[118] Montalbetti E, Biscéré T, Ferrier-Pagès C, et al. Manganese benefits heat-stressed corals at the cellular level. Frontiers in Marine Science. 2021;**8**:1-14, 681119. DOI: 10.3389/fmars.2021.681119]

[119] Chow AM, Steel R, Anderson RL. Hsp72 chaperone function is dispensable for protection against stress-induced apoptosis. Cell Stress & Chaperones. 2009;**14**(3):253-263. DOI: 10.1007/S12192-008-0079-4/FIGURES/5

[120] Seveso D, Montano S, Maggioni D, et al. Diel modulation of Hsp70 and Hsp60 in corals living in a shallow reef. Coral Reefs. 2018;**37**(3):801-806. DOI: 10.1007/S00338-018-1703-0/FIGURES/4

[121] Downs CA, Kramarsky-Winter E, Woodley CM, et al. Cellular pathology and histopathology of hypo-salinity exposure on the coral Stylophora pistillata. Science of the Total Environment. 2009;**407**(17):4838-4851. DOI: 10.1016/J.SCITOTENV.2009.05.015

[122] Seveso D, Montano S, Strona G, Orlandi I, Galli P, Vai M. Exploring the effect of salinity changes on the levels of Hsp60 in the tropical coral Seriatopora caliendrum. Marine Environmental Research. 2013;**90**:96-103. DOI: 10.1016/J.MARENVRES.2013.06.002

[123] Thummasan M, Casareto BE, Ramphul C, Suzuki T, Toyoda K, Suzuki Y. Physiological responses (Hsps 60 and 32, caspase 3, $H_2O_2$ scavenging, and photosynthetic activity) of the coral *Pocillopora damicornis* under thermal and

high nitrate stresses. Marine Pollution Bulletin. 2021;**171**:1-10, 112737. DOI: 10.1016/j.marpolbul.2021.112737

[124] Seveso D, Montano S, Strona G, Orlandi I, Vai M, Galli P. Up-regulation of Hsp60 in response to skeleton eroding band disease but not by algal overgrowth in the scleractinian coral Acropora muricata. Marine Environmental Research. 2012;**78**:34-39. DOI: 10.1016/J.MARENVRES.2012.03.008

[125] Milne Edwards H. Histoire naturelle des coralliaires ou polypes proprement dits 3: 1-560. Paris: Librairie Encyclopédique de Roret; 1860

[126] Dana JD. Zoophytes. United States Exploring Expedition During the Years 1838-1842. Vol. 7. Philadelphia: Lea and Blanchard; 1846. pp. 1-740. 61 pls. (1846: 1-120, 709-720; 1848: 121-708, 721-740; 1849: atlas pls. 1-61). Available online at: http://www.sil.si.edu/digitalcollections/usexex/navigation/ScientificText/USExEx19_08select.cfm

[127] Barshis DJ, Ladner JT, Oliver TA, Seneca FO, Traylor-Knowles N, Palumbi SR. Genomic basis for coral resilience to climate change. Proceedings of the National Academy of Sciences of the United States of America. 2013;**110**(4):1387-1392. DOI: 10.1073/PNAS.1210224110/-/DCSUPPLEMENTAL

[128] de Lamarck J-BM. Histoire naturelle des animaux sans vertèbres. Tome second. Paris: Verdière; 1816. 568 pp. Available online at: http://www.biodiversitylibrary.org/item/47698

[129] Thompson DM, Van Woesik R. Corals escape bleaching in regions that recently and historically experienced frequent thermal stress. Proceedings of the Royal Society B: Biological Sciences. 2009;**276**(1669):2893-2901. DOI: 10.1098/RSPB.2009.0591

[130] Oliver TA, Palumbi SR. Do fluctuating temperature environments elevate coral thermal tolerance? Coral Reefs. 2011;**30**(2):429-440. DOI: 10.1007/s00338-011-0721-y

[131] Thomas L, Rose NH, Bay RA, et al. Mechanisms of thermal tolerance in reef-building corals across a Fine-grained environmental mosaic: Lessons from Ofu, American Samoa. Frontiers in Marine Science. 2018;**4**:434. DOI: 10.3389/FMARS.2017.00434/BIBTEX

[132] Angilletta MJ. Estimating and comparing thermal performance curves. Journal of Thermal Biology. 2006;**31**(7):541-545. DOI: 10.1016/j.jtherbio.2006.06.002

[133] McIlroy SE, Thompson PD, Yuan FL, Bonebrake TC, Baker DM. Subtropical thermal variation supports persistence of corals but limits productivity of coral reefs. Proceedings of the Royal Society B: Biological Sciences. 2019;**286**(1907):20190882. DOI: 10.1098/rspb.2019.0882

[134] Bairos-Novak KR, Hoogenboom MO, van Oppen MJH, Connolly SR. Coral adaptation to climate change: Meta-analysis reveals high heritability across multiple traits. Global Change Biology. 2021;**27**(22):5694-5710. DOI: 10.1111/GCB.15829

[135] Costanza R, de Groot R, Sutton P, et al. Changes in the global value of ecosystem services. Global Environmental Change. 2014;**26**(1):152-158. DOI: 10.1016/j.gloenvcha.2014.04.002

[136] Kleypas JA, McManus JW, Menez LAA. Environmental limits to coral reef development: Where do we draw the line? American Zoologist. 1999;**39**(1):146-159. Available from: https://academic.oup.com/icb/

article/39/1/146/124572 [Accessed: 06-07-2020]

[137] Perry CT, Larcombe P. Marginal and non-reef-building coral environments. Coral Reefs. 2003;**22**:427-432. DOI: 10.1007/s00338-003-0330-5

[138] Dellisanti W, Tsang RHLRHL, Ang P, Wu J, Wells MLML, Chan LLLL. Metabolic performance and thermal and salinity tolerance of the coral Platygyra carnosa in Hong Kong waters. Marine Pollution Bulletin. 2020;**153**:111005. Available from: https://www.sciencedirect.com/science/article/pii/S0025326X20301235?dgcid=author [Accessed: 02-03-2020]

[139] Fisher R, Bessell-Browne P, Jones R. Synergistic and antagonistic impacts of suspended sediments and thermal stress on corals. Nature Communications. 2019;**10**(1):1-9. DOI: 10.1038/s41467-019-10288-9

[140] Kavousi J, Keppel G. Clarifying the concept of climate change refugia for coral reefs. ICES Journal of Marine Science. 2018;**75**(1):43-49. DOI: 10.1093/ICESJMS/FSX124

[141] Duprey NN, McIlroy SE, Ng TPT, et al. Facing a wicked problem with optimism: Issues and priorities for coral conservation in Hong Kong. Biodiversity and Conservation. 2017;**26**(11):2521-2545. DOI: 10.1007/s10531-017-1383-z

[142] de Soares MO. Marginal reef paradox: A possible refuge from environmental changes? Ocean and Coastal Management. 2020;**185**:105063. DOI: 10.1016/J.OCECOAMAN.2019.105063

[143] Dixon AM, Forster PM, Heron SF, Stoner AMK, Beger M. Future loss of local-scale thermal refugia in coral reef ecosystems. PLOS Climate. 2022;**1**(2):e0000004. DOI: 10.1371/JOURNAL.PCLM.0000004

[144] Kroeker KJ, Kordas RL, Crim RN, Singh GG. Meta-analysis reveals negative yet variable effects of ocean acidification on marine organisms. Ecology Letters. 2010;**13**(11):1419-1434. DOI: 10.1111/j.1461-0248.2010.01518.x

[145] Darling ES, D'agata S. Coral reefs: Fishing for sustainability. Current Biology. 2017;**27**(2):R65-R68. DOI: 10.1016/J.CUB.2016.12.005

[146] McClanahan TR. Marine reserve more sustainable than gear restriction in maintaining long-term coral reef fisheries yields. Marine Policy. 2021;**128**:104478. DOI: 10.1016/J.MARPOL.2021.104478

Chapter 2

# The Mauritanian Slope (NE Atlantic) Has No Desert: *Swiftia phaeton* (Holaxonia: Plexauridae) Shaping Coral Gardens

*Íris Sampaio, Lydia Beuck, Gui M. Menezes and André Freiwald*

## Abstract

*Swiftia* Duchassaing & Michelotti, 1864, is often found sparse in the NE Atlantic Ocean and Mediterranean Sea. When the cruise MSM 16/3 "PHAETON" filmed the upper bathyal off Mauritania in 2010, the first dense populations dominated by *Swiftia* were discovered in the NE Atlantic Ocean, co-occurring with the framework-forming scleractinians *Desmophyllum pertusum* (Linnaeus, 1758) and *Madrepora oculata* Linnaeus, 1758. Remotely operated vehicle (ROV) video annotation from two canyons and two coral mounds considered *Swiftia phaeton* Sampaio, Beuck & Freiwald, 2022 presence, size class and abundance as well as substrate and geomorphology of the seafloor. Coral gardens definition included abundance and size of the species. Dense and very dense mono- and multispecific coral gardens dominated by *S. phaeton* were mapped between 20°24′N and 17°54′N in 470–640 m depth. The resilience of these coral gardens off Mauritania is mainly linked to the presence of hard substrate available to settle and to the exposition of currents rich in food. Still, these ecosystems are located inside a hydrocarbon exploration area off Mauritania and where fisheries occur since the 1960s. Sedimentation plumes caused by both activities can travel and impact on settlement, development, and survival of these populations. Hence, these vulnerable "oases" should be protected.

**Keywords:** canyons, conservation, coral mounds, deep sea, NW Africa, Octocorallia, vulnerable marine ecosystem (VME)

## 1. Introduction

The contemporary era of deep-sea exploration is related to the exponential development of the technology able to map, to visualize, and to sample extreme environments. In consequence, coral ecosystems are being discovered at a higher rate than ever with remotely operated vehicles (ROVs) in unexplored deep-sea areas [1–4]. Octocorals are the most dominant subclass of corals in cold waters of deep-sea environments, forming high-density assemblages with the dominance of a single species or multispecific coral gardens in the NE Atlantic Ocean [5–7]. As key elements of coral gardens, octocorals are ecosystem engineers that create three-dimensionality for feeding and spawning, and to serve as nursery grounds for a great range of invertebrate taxa and fish species [5, 8, 9]. Though representing a prodigious amount of biodiversity, octocorals are characterized by life traits such as long life span, slow growth, low fecundity, late maturity, and reproduction, which translate in low resilience and slow recovery from the human activities. Major anthropogenic impacts on octocorals include bottom trawling, seabed mining, and hydrocarbon drilling, but natural geological processes such as landslides can also be threats to octocorals [10–13].

Academia awareness and call for urgent action on the unregulated exploitation of cold-water coral and sponge ecosystems have led the United Nations to act [14]. The United Nations General Assembly [15] firstly created a moratorium on bottom trawling at High Seas (Resolution 61/105) [16]. Then, it called the attention of governments and regional organizations for research, mapping, and conservation ecosystems from bathyal cold waters [16, 17]. The Food and Agriculture Organization (FAO) followed, defining the management standards and criteria for the definition of Vulnerable Marine Ecosystems (VMEs) [18]. Henceforward, the Convention for the Protection of the Marine Environment of the North-East Atlantic has considered coral gardens, including octocorals as the key components of VMEs [19, 20]. Now, Europe is leader on deep-sea habitat classification schemes of the NE Atlantic Ocean, which have their southern threshold at the Strait of Gibraltar. Hence, both European Nature Information System (EUNIS) and the CoraFISH cold-water coral biotope classification scheme do not consider African NE Atlantic VMEs, which are excluded from further management actions [21, 22].

The North Atlantic Ocean is the most explored oceanic area in the world. Yet, the NW of Africa's remains understudied. The deep sea of NW Africa has been explored only since the late nineteenth century. Here exploration occurred sporadically, for example, during the Talisman and *Michael Sars* North Atlantic Deep-Sea expedition (1883 and 1910, respectively) and regularly by later Dutch and Spanish expeditions [23–27]. Nowadays, exploration targets also nearby Areas Beyond National Jurisdiction (ABNJ) [28].

Few scientific expeditions have specifically focused at the Mauritanian coast (*Tydeman* Madeira-Mauritania—CANCAP III in 1978, *Tyro* Mauritania II in 1988 and Spanish MAURIT surveys (2007–2010)) [24, 25, 27]. Yet, at Mauritania is located the world's largest coral mound barrier, parallel to the coast [11, 29]. While deep-sea benthonic fauna from Mauritania is mostly unknown, live scleractinians were already found sparsely distributed through coral mounds, and widespread scleractinian framework was discovered in canyons off Mauritania [30–32]. Few octocorals, including *Swiftia dubia* (Thomson, 1929) [104], were already reported to the deep sea of the region by the Maurit-Expeditions [27].

Octocoral gardens seem to be widespread throughout the NE Atlantic Ocean from the Hardangerfjord in Norway to the Mid-Atlantic Ridge (MAR) and island slopes and seamounts of the Azores; and in the Mediterranean Sea [5, 7, 33–35]. In this area, colonies of three species of the genus *Swiftia* Duchassaing and Michelotti, 1864 [91] were mostly found to be sparsely distributed and associated with hard substrates, between 20 and 2400 m depth [36, 37]. Nonetheless, *Swiftia dubia* (as *S. pallida* Madsen, 1970 [101]) is considered a dominant species of the "cold-water gorgonians on hard/mixed substrate or compact mud" category of the CoralFISH cold-water coral biotope classification scheme, indicating the presence of *Swiftia* coral gardens in the area [22].

In 2010, RV *Maria S. Merian* cruise MSM 16/3 "PHAETON"—Paleoceanographic and paleoclimatic record on the Mauritanian shelf off Mauritania was targeting the coral-mound barrier off Mauritania with an ROV and an exploratory mindset that led to the discovery of octocoral gardens throughout its slope [30]. As a result of this expedition, a new species of the genus *Swiftia* was discovered and recently submitted on a distinct publication including its distribution [38]. With this study, we aim to characterize the first African biotope of the genus *Swiftia* in the NE Atlantic Ocean, to discuss its distribution, considering biotic factors and anthropogenic influences, using it as a case study to discuss the definition of the concept "coral garden" and its implications for deep-sea conservation.

## 2. Materials and methods

### 2.1 Study area

The NW Africa, offshore Mauritania, is where a coral mound province was mapped from the Tanoûdêrt Canyon at 20°15′N until close to the Senegalese border at 16°03′N (**Figure 1**) [39]. Through at least 580 km, the coral mounds represent one of the largest known complexes positioned in two-slope parallel chains and named Mauritanian Coral Province [29, 32]. The coral mounds are on the bathymetrical range between 400 and 550 m depth and can achieve a height up to 100 m, supporting abundant associated species and live corals [30]. New species associated with this coral mound complex have been described recently: Bryozoa [40], Hydrozoa [41], Porifera [42], Crustacea [43], and Mollusca [44]. The live reef-forming corals associated with the coral mound complex of Mauritania are the abundant *Desmophyllum pertusum* (Linnaeus, 1758) [100] and *Madrepora oculata (*Linnaeus, 1758) [100] also associated with other scleractinians such as *Dendrophyllia cornigera* (Lamarck, 1816) [99], *Solenosmilia variabilis* Duncan, 1873 [92], and *Desmophyllum dianthus* (Esper, 1794) [45, 94].

At the same time, several gullies divide the slope and merge into submarine canyon systems that may extend over several hundreds of kilometers deep off Mauritania [39]. While corals are scarce on Mauritanian mounds, canyons have abundant corals and may act as refuges that permit recolonization of the mounds [32].

Mauritania is an example of the existence of eutrophic large-scale tropical ecosystems with a major coastal upwelling system that leads to high-productivity waters [46]. There, oceanography is influenced by intermediate water masses: the North and South Atlantic Central Waters (NACW and SACW) from 150 to 600 m depth, the Antarctic Intermediate Water (AAIW) from 600 to 1000 m depth, and the North Atlantic Deep Water (NADW) from 1000 to 1600 m depth [47, 48]. Besides, the Poleward undercurrent (PUC) that runs along the slope between 100 and 300 m depth carries the SACW to the north [49]. Both the SACW and the NACW met off Cape Blanc and are forced seaward to form the Cape-Verde Frontal Zone with permanent upwelling conditions north of the oceanic front and seasonal upwelling, south of the oceanic front.

From the latitude of 5°N–20°N, there is an Oxygen Minimum Zone (OMZ) (~300–700 m depth) at the SACW lower core, above the AAIW [50]. The upwelling of nutrient-rich waters occurring in this area generates high productivity on the surface, which by export of high organic matter and its degradation leads to higher consumption of oxygen [32].

In 2010, the cruise MSM 16/3 "PHAETON"—Paleoceanographic and paleoclimatic record on the Mauritanian shelf off Mauritania onboard RV *Maria S. Merian* targeted

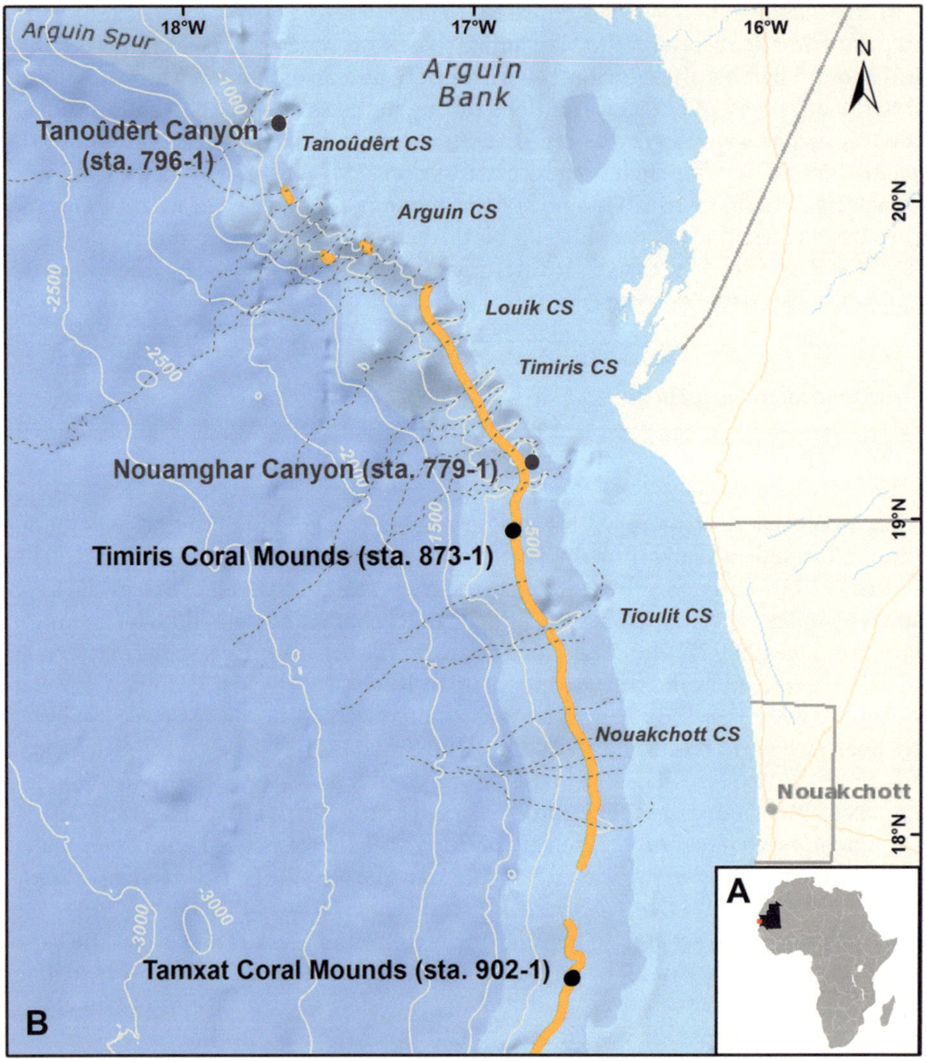


**Figure 1.**
*Location of (A) Mauritania and (B) the four ROV dives of MSM 16/3—Phaeton on the continental slope off Mauritania analyzed herein. Highlighted stations with site name and GeoB station (sta.): grey = canyons; black = coral mounds; orange = scleractinian distribution. Basemap from ESRI [85], contours from GEBCO compilation group [86], scleractinian distribution, canyon position and names from [39].*

the upper canyon flanks and coral mound complexes along the Mauritanian margin, where footage of octocorals was collected (**Figure 1**) [30].

## 2.2 ROV survey

Twelve dives were performed by an ROV along an N-S stretch of about 353 km. The spatial coverage of the dives went from off Cape Blanc (20°14′N) to the south (17°08′N) and the longitude 16°39′W to 17°40′W, between 417 m and 642 m depth on the continental slope off Mauritania (**Figure 1**; **Table 1**). Vessel-based Posidonia system was used as positioning system, and *via* time code, the video footage was linked to the navigation track. Footage was collected by the ROV Sperre AS

*DOI: http://dx.doi.org/10.5772/intechopen.104635*

| Dive No. | Area | Station | Latitude | Longitude | Depth (m) | | DO (m/l) | S (PSU) | T (°C) |
|---|---|---|---|---|---|---|---|---|---|
| | | | | | Shallower | Deeper | | | |
| 1 | Arguin south canyon | 14.759–1 | 19°44′03″ to 19°44′16″ | −17°08′44″ to −17°08′50″ | 488 | 546 | | | |
| 2 | Nouamghar canyon | 14.779–1 | 19°10′47″ to 19°10′36″ | −16°48′21″ to −16°48′17″ | 449 | 619 | 1.54 | 35.21 | 9.95 |
| 3 | Tanoûdêrt canyon | 14.796–1 | 20°14′50″ to 20°14′35″ | −17°40′12″ to −17°40′04″ | 487 | 642 | 1.63 | 35.20 | 9.95 |
| 4A | Inchiri canyon | 14.871–1 | 19°08′21″ to 19°08′22″ | −16°45′53″ to −16°45′49″ | 519 | 589 | | | |
| 4B | Inchiri canyon | 14.871–2 | 19°08′21″ to 19°08′14″ | −16°45′51″ to −16°45′40″ | 427 | 564 | | | |
| 5 | Deep Timiris mound complex | 14.873–1 | 18°57′41″ to 18°57′53″ | −16°52′17″ to −16°52′01″ | 480 | 603 | 1.39 | 35.27 | 10.63 |
| 6 | Shallow Timiris mound complex | 14.874–1 | 18°58′00″ to 18°57′36″ | −16°51′15″ to −16°51′04″ | 429 | 525 | | | |
| 7 | Tioulit canyon (S) | 14.886–1 | 18°39′01″ to 18°38′29″ | −16°43′35″ to −16°43′45″ | 475 | 641 | | | |
| 8 | Tioulit canyon (N) | 14.891–1 | 18°39′51″ to 18°39′57″ | −16°43′26″ to −16°43′29″ | 502 | 592 | | | |
| 9 | Tamxat mound complex (c) | 14.902–1 | 17°32′28″ to 17°32′51″ | −16°40′06″ to −16°39′41″ | 396 | 588 | 1.20 | 35.27 | 10.80 |
| 10 | Banda mound complex | 14.908 | 17°40′13″ to 17°40′12″ | −16°40′50″ to −16°40′17″ | 455 | 574 | | | |
| 11 | Tamxat mound complex (S) | 14.909–1 | 17°28′57″ to 17°28′57″ | −16°41′57″ to −16°41′28″ | 423 | 560 | | | |
| 12 | Tiguent mound complex | 14.914 | 17°08′12″ to 17°07′54″ | −16°49′29″ to −16°48′53″ | 409 | 515 | | | |

**Table 1.**
*ROV dives performed during MSM 16/3—Phaeton on the shelf and continental slope off Mauritania. Details: Dive number, area where the dive took place, station number, latitude, longitude, depth range, and mean oceanographic variables at coral garden locations (mean oxygen concentration, mean salinity, and mean temperature).*

Sub-fighter 7500 DC (Sven Lovén Centre for Marine Infrastructure, University of Gothenburg) on board the R/V Maria S. Merian. The ROV equipment encompassed a Sperre HD video camera (1080 I and 720 p), two standard video cameras and a still camera (Canon Powershot G9, 12 Mpixel), two Deep Sea Systems red lasers (50 mm distance), and a HYDRI-LEK-5-function hydraulic manipulator-type EH and sampler box [30]. A total video record of 47 h 83 m totalized ~3000 still images with 12-Mp resolution. During the dive, the video signal from the front-looking camera was stored in digital Quick Time MPEG-4 format (*.mov) with a resolution of 1280x720 pixels, a bit rate of 100 MB/s, and in sequences of about 5 minutes each. Oxygen concentration (ml/l), salinity (PSU), and temperature (°C) were measured with ROV-mounted sensors. Cleaning of the ROV navigation track was done with the ArcGIS extension Adelie of IFREMER by manual cleaning and Gaussian smoothing.

### 2.3 Video annotation of *Swiftia phaeton*

The presence of *Swiftia phaeton* Sampaio, Beuck & Freiwald 2022 [42] was scrutinized through four ROV dives, where it occurred in higher abundance, in particular two dives in canyons (Nouamghar and Tanoûdêrt canyons) and two dives in coral mound complexes (deep Timiris Mound Complex and Tamxat Mound Complex). High-resolution videos of the four dives were annotated every 10 seconds considering the following aspects: *S. phaeton* presence, size class, and relative abundance. ROV footage lacked fundamental variables to calculate size of the colonies and area of the footage. Therefore, three size classes of colonies of *S. phaeton* were considered: 1) small (<5 cm in length), 2) medium (5–15 cm in length), and 3) large (15–30 cm in length) (**Figure 2A–C**). The relative abundance of *S. phaeton* was estimated through the establishment of a length/distance ratio between individuals of the same species, divided into five classes: 1) isolated (> 15 m = >50 lengths of distance); 2) scattered (150 cm–15 m = 5–50 lengths of distance); 3) frequent (90–150 m = 3–5 lengths of distance); 4) dense (30–90 cm = 1–3 lengths of distance); and 5) very dense (<30 cm = <1 length of distance) (**Figure 2D–H**).

Coral gardens were defined and identified based on the two classes of abundance, dense and very dense, and their horizontal and vertical distributions were mapped. In order to understand which factors could be influencing the distribution of the species and the coral gardens it forms, we annotated the type of substrate and the geomorphology of the seafloor. A dataset with the complete video annotation was deposited at the World Data Center Pangaea (https://doi.pangaea.de/10.1594/PANGAEA.910893). Finally, the main components of the megabenthic community and *S. phaeton* associated fauna were identified to the lowest possible taxonomic level.

Maps were based on the most accurate bathymetrical data, collected by the ROV.

Analyses of the data, done using the Excel plug-in XLStat 2019.1 (Addinsoft, Boston, USA), excluded annotations in which the ROV was not in movement to avoid repetition. Three datasets were then considered for comparison: (1) absence and presence of *S. phaeton*, (2) presence of *S. phaeton*, and (3) presence of coral gardens formed by *S. phaeton*. Multiple correspondence analyses (MCA) were performed to understand how factors were linked to the distribution of the species and to the distribution of the coral gardens this species forms. Chi-squared tests were made to verify the link between pairs of two factors: (1) substrate and size, (2) type of location and size, (3) geomorphology and size, and (4) depth and type of location, on the presence of the species and the coral gardens it forms.

**Figure 2.**
*Size and relative abundance classes of* Swiftia phaeton *considered in the video annotation of four ROV dives performed during MSM 16/3—Phaeton off Mauritania: (A) small, (B) medium, and (C) large colonies and (D) isolated, (E) scattered, (F) frequent, (G) dense, and (H) very dense colonies and communities. Types of coral gardens formed by* Swiftia phaeton *are also represented in (C) monospecific coral garden and (G) multispecific coral garden with undescribed Plexauridae species.*

## 3. Results

### 3.1 Distribution of *Swiftia phaeton*

Detailed maps of *S. phaeton* distribution were made based on video annotation of Nouamghar and Tanoûdêrt Canyons and deep Timiris and Tamxat Mound Complexes (**Figure 3**; **Table 1**). The species is widespread at Nouamghar Canyon and deep

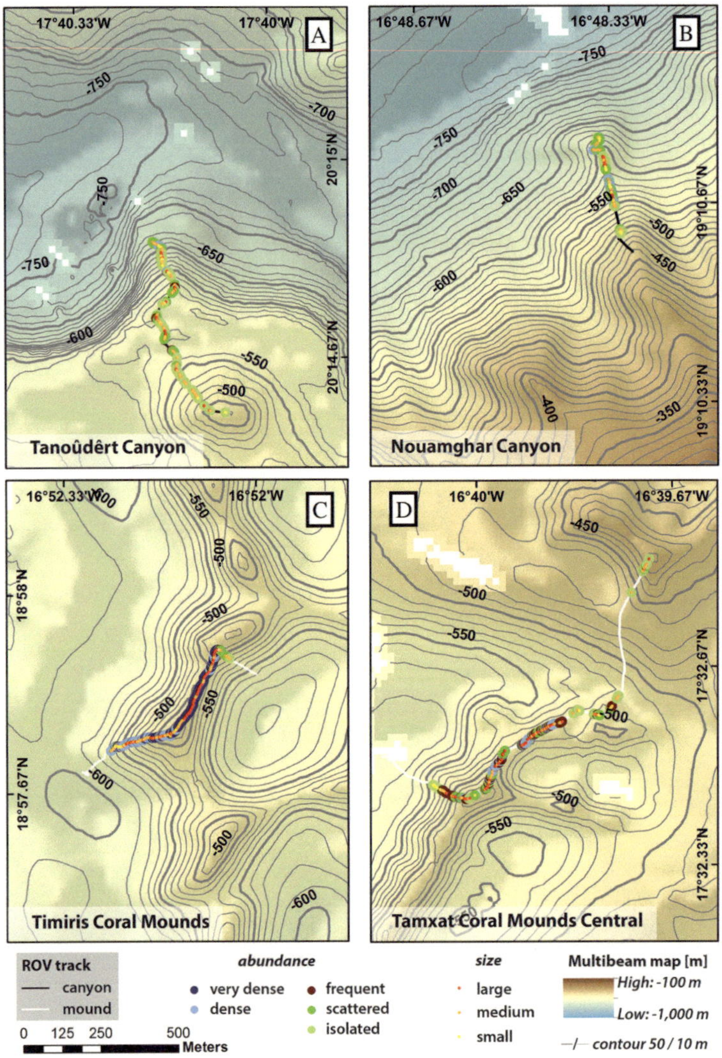


**Figure 3.**
*Distribution of Swiftia phaeton and the coral gardens it forms along four locations off Mauritania where it is more common. Video annotation of ROV dive tracks has considered the species size and relative abundance. Dense and very dense abundances characterize coral gardens. Bathymetry from [51].*

Timiris Mound Complex, yet it is absent at specific depths. At Tanoûdêrt Canyon, *S. phaeton* is also widespread but mainly sparsely distributed. Its deepest record is localized at 550 m depth, on the deeper coral mound of the central Tamxat Mound Complex.

Presence of *S. phaeton* at Mauritania varies from isolated colonies to very dense coral gardens, dominating the monospecific biotope or in association with other Plexauridae species forming multispecific biotopes (**Figure 3**). Herein, we report the first coral gardens dominated by a species of the genus *Swiftia* on the NE Atlantic Ocean (**Figure 3**). The extensive habitats dominated by *S. phaeton* were recorded between 20°24′N and 17°54′N in 470–640 m depth, co-occurring with the framework-forming scleractinians *Desmophyllum pertusum* (Linnaeus, 1758) [100] and *Madrepora oculata* Linnaeus, 1758 [100] (**Figure 3**). Coral gardens of *S. phaeton* are predominant at the Nouamghar Canyon and at the Timiris Mound Complex, though also occurring at the Tanoûdêrt Canyon and the Tamxat Mound Complex (**Figure 3**). These coral gardens are mostly formed by dense aggregations of branched, large (15–30 cm in length), or medium (5–15 cm in length) colonies of *S. phaeton* and some very dense assemblages (**Figures 2B, C, H** and **3**).

Dense coral gardens were observed scattered at the flank of Nouamghar Canyon: from 520 to 530 m and from 610 to 620 m depth between 19°10′47″N and 19°10′37″W and between 16°48′21″W and 16°48′19″W, respectively (**Figures 3** and **4**). There, the *S. phaeton* garden has the largest extension of ~523 m along the ROV track, and it is located between 570 and 595 m depth with medium-sized colonies and from 570 to 580 m depth with larger colonies (**Figures 3** and **4**). The dense coral gardens of the Tanoûdêrt Canyon were found between 20°14′47″N and 20°14′51″N and between 17°40′10″W and 17°40′11″W. There, the coral garden with the longest lateral extension along the ROV track is formed by larger colonies between 565 and 590 m depth but the canyon flank also has two coral gardens formed by small (5 cm in height) colonies between 630 and 640 m depth (**Figures 3** and **4**).

The most extended coral garden of all the study sites was recorded at the eastern mound top and the ridge, between two mounds of the deep Timiris Mound Complex, from 18°57′44″N to 18°57′54″N and from 16°52′15″W to 16°52′04″W and between

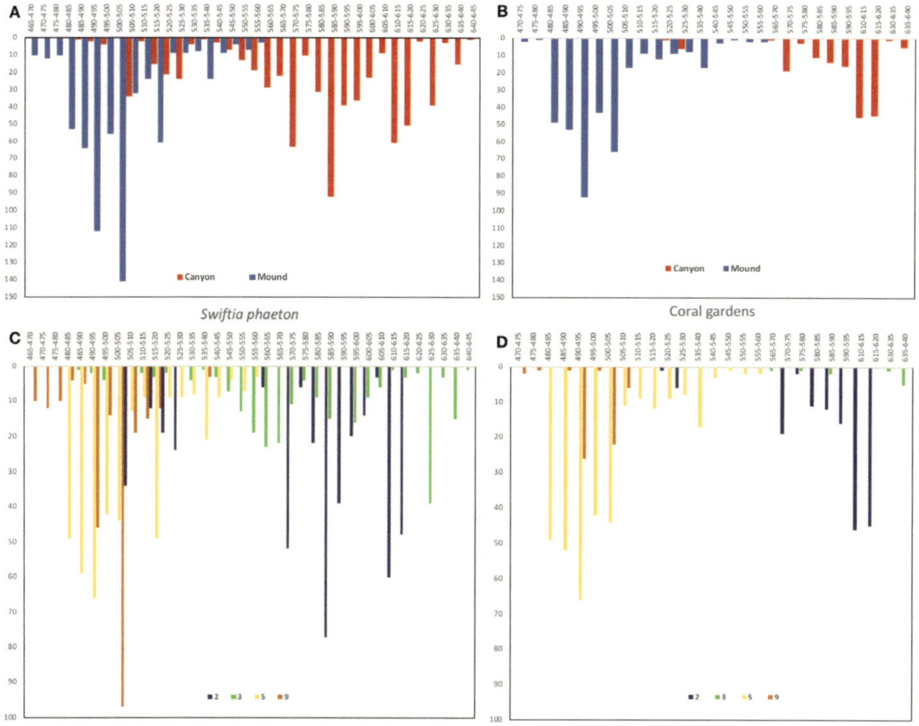


**Figure 4.**
*Bathymetric distribution of Swiftia phaeton and the coral gardens it forms off Mauritania, at distinct types of locations (A, B) and at different locations (ROV dives) (C, D).*

480 and 560 m depth (**Figures 3** and **4**). Here, very dense aggregations of *S. phaeton* were recorded dominated by large colonies. Besides, there are two coral mounds with a widespread coral garden starting at 560 m depth on the western flank of one and ending on the second mound top. At the central Tamxat Mound Complex, dense coral gardens with large colonies were located sparse at the western flank of the deeper coral mound and at the ridge between coral mounds, from 17°32′27″N to 17°32′35″N and from 16°40′03″W to 16°39′48″W between 470 and 510 m depth (**Figures 3** and **4**).

The bathymetric distribution of the coral gardens formed by *S. phaeton* is mostly focused on the range of 490–495 m depth. Yet, coral gardens are also frequent between 480 and 505 m depth and between 610 and 620 m depth (**Figure 4**). These ecosystems are significantly deeper when found in canyons (520–640 m depth), when compared with mound locations (470–510 m depth) (chi-square and Fisher's exact test <0.0001) (**Figure 4A**, **B**).

## 3.2 Factors influencing *Swiftia phaeton*'s distribution

The distributions of the species and the coral gardens it forms are influenced by several factors, such as substrate and geomorphology. At Mauritania, distinct variables of each factor seem to be related to specific locations, as revealed by the results of two MCAs on the presence of the species and of the coral gardens (**Figure 5**).

The Tanoûdêrt Canyon has isolated and scattered colonies that grow on framework located at the canyon shoulder/plateau (**Figures 3** and **5A**). Its coral gardens have dense assemblages living on framework (**Figures 3** and **5B**). At Nouamghar Canyon, the species is small or medium in size, it prefers rocky substrate, and it lives on the coral mound or at the canyon flank, while the coral gardens only exist at the canyon flank (**Figures 3** and 5). Very dense communities, which settle on coral rubble at the mound flank or at the mound top, characterize the Timiris Mound Complex, whereas large colonies inhabit the ridge, either in frequent aggregations or forming coral gardens at the Tamxat Mound Complex (**Figure 5**).

### *3.2.1 Type of location*

The species is present in both canyons and on coral mounds, but the coral gardens are mostly found on mounds. Thus, the type of location was significantly related to the size of the species, even when it forms coral gardens (chi-square and Fisher's exact test <0.0001). While larger colonies predominantly inhabit coral mounds, medium or small ones dominate in canyons (**Figure 6A**). Though, when forming coral gardens, a lower proportion of smaller colonies exist on canyons. Likewise, a higher proportion of larger and medium colonies exist in coral mounds (**Figure 6B**).

### *3.2.2 Substrate*

Higher abundance of *S. phaeton* was exclusively found when hard substrate (coral framework, coral rubble, and rocky substrate) was present, while the absence of the species dominated with soft sediment. Type of substrate was found to be significantly related to the size of the colonies, even when forming coral gardens (chi-square and Fisher's exact test <0.0001). While the colonies settling on coral rubble were medium or large, when forming a coral garden, the colonies were mostly large sized (**Figure 6C** and **D**). On the other hand, when the colonies live

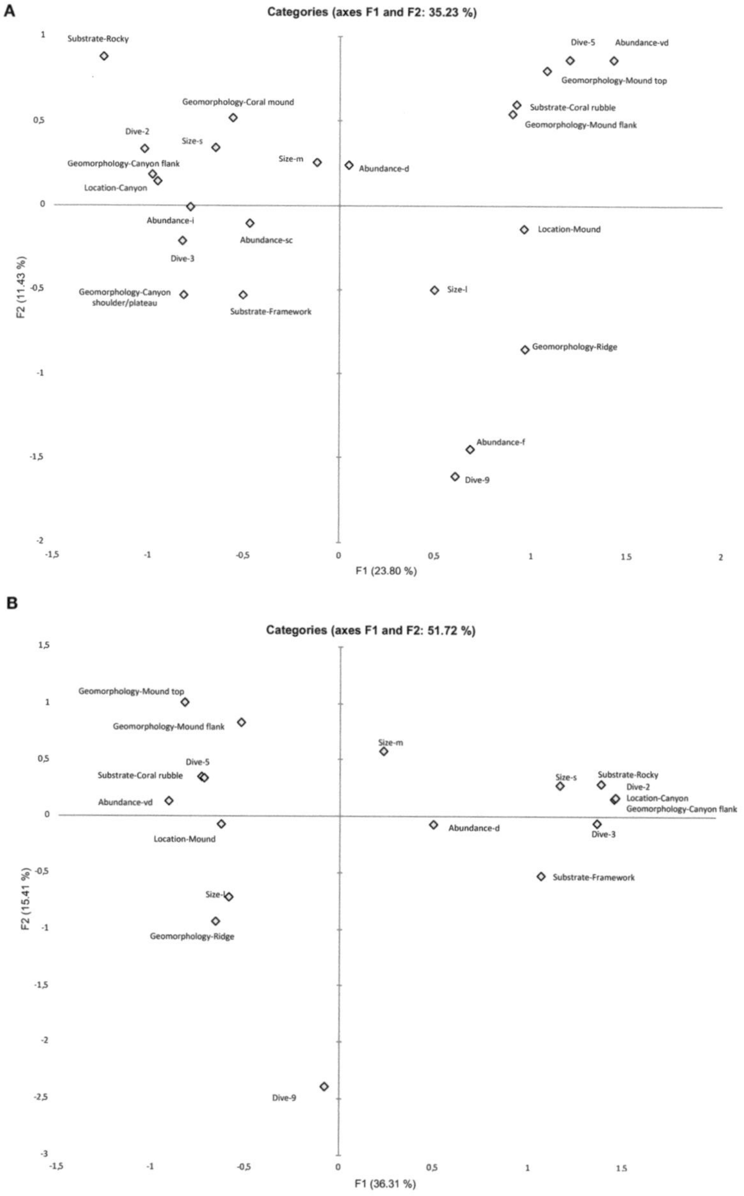


**Figure 5.**
*Multiple correspondence analysis (MCA) on the distribution of* Swiftia phaeton *(A) and on the distribution of coral gardens formed by* Swiftia phaeton *off Mauritania (B).*

on scleractinian framework, they have medium size, especially where coral gardens exist. Rocky substrate supports medium colonies of the species or large colonies, when it forms coral gardens (**Figure 6C** and **D**).

### *3.2.3 Geomorphology*

Annotations revealed that *S. phaeton* prevails at canyon flanks and ridges, while it is absent on a slope (Tamxat coral mounds), a coral mound, and at mound bases.

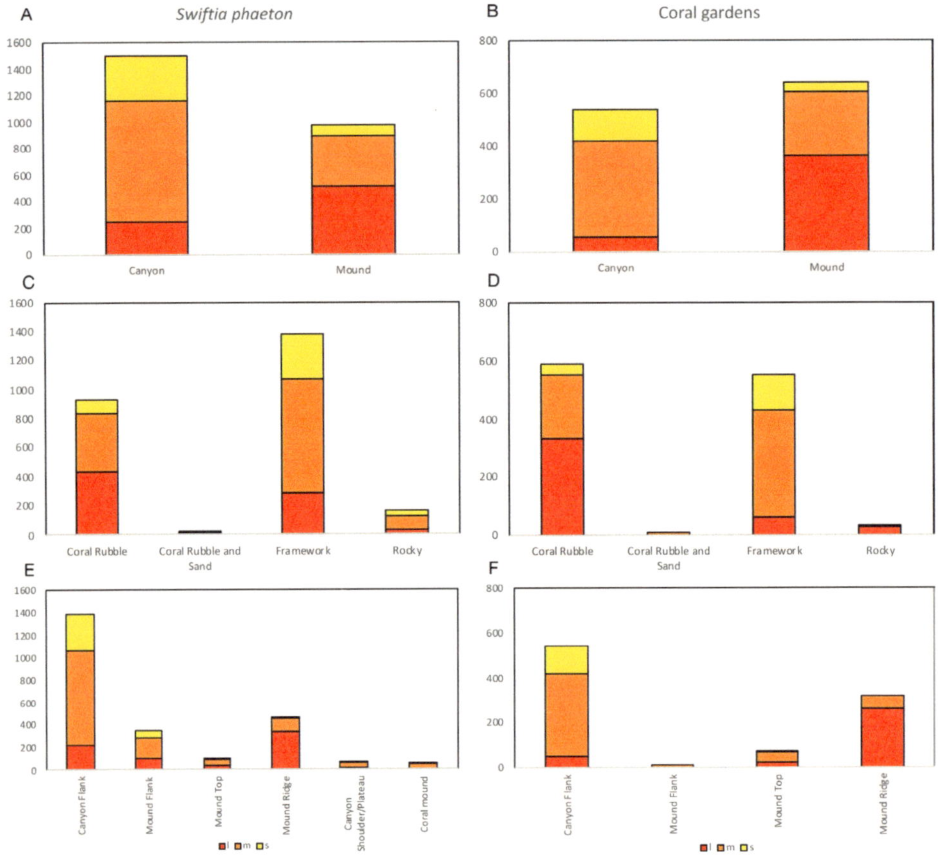


**Figure 6.**
*Distribution of the size of colonies of Swiftia phaeton (l: large, m: medium, s: small), in general and when forming coral gardens at four selected locations off Mauritania according to the influence of distinct factors: type of location (A, B), type of substrate (C, D), and geomorphology (E, F).*

Geomorphological features of the deep sea off Mauritania were discovered to be significantly related to the size of the colonies of *S. phaeton*, and with the size of the colonies when this species forms coral gardens (chi-square and Fisher's exact test <0.0001).

Larger colonies of *S. phaeton* were found associated with ridges between coral mounds, while smaller colonies were mainly found at flanks of canyons and mounds (**Figure 6E**). The same occurred with larger colonies forming coral gardens, but the smaller colonies of coral gardens were found at flanks of canyons. Nonetheless, most colonies forming coral gardens, when associated with canyon flanks, had medium size (**Figure 6F**).

### 3.3 Fauna associated with *Swiftia phaeton*'s coral gardens

Monospecific coral gardens formed by *Swiftia phaeton* are able to agglomerate diverse associated fauna. The fauna associated with *S. phaeton*'s coral gardens is mainly constituted by crabs such as *Eumunida bella* de Saint-Laurent and Macpherson, 1990 [88] (**Figure 7F**) and *Paromola cuvieri* (Risso, 1816) [102] but also

**Figure 7.**
Swiftia phaeton *coral gardens community and associated fauna: (A)* Swiftia phaeton *and* Thesea talismani *Grasshoff, 1986 [95] on* Desmophyllum pertusum *(Linnaeus, 1758) [100] and* Acesta *sp. frameworks, (B)* Desmophyllum pertusum *with associated* Cladorhiza corallophila *Göcke, Hestetun, Uhlir, Freiwald, Beuck & Janussen, 2016 [42], (C) undescribed Plexauridae, (D)* Thesea talismani *with associated zoantharian, (E) cf.* Clavularia borealis *Koren and Danielssen, 1883 [98] on coral rubble, (F)* Paromola cuvieri *(Risso, 1816) [102] with Plexauridae, (G) "Lithistidae", (H)* Helicolenus dactylopterus *(Delaroche, 1809) [89], (I)* Pteroctopus tetracirrhus *(Delle Chiaje, 1830) [90], and (J)* S. phaeton *associated eggcase.*

by a diverse variety of animals such as sponges, fish, octopus, mollusks, and grazing echinoderms such as seastars (**Figures 2** and 7). Grazing behavior of mollusks was also observed in some colonies of this species (see also [52]).

When *S. phaeton*'s coral gardens are part of multispecific biotopes, the upper bathyal of Mauritania's megabenthic community has a patchy distribution of other species of octocorals from the families Plexauridae Gray, 1859 [96], Anthothelidae

Broch, 1916 [87], and Acanthogorgiidae Gray, 1859 [96], but also from the suborder Stolonifera Thomson and Simpson, 1909 [103] (**Figure** 7).

## 4. Discussion

*Swiftia phaeton* Sampaio, Beuck & Freiwald, 2022 [42] is the first species of its genus to be recorded as a fundamental habitat builder of coral gardens. Coral gardens formed by *S. phaeton* were discovered predominantly in mounds of the Mauritanian slope. The most extensive coral garden was found at the deep Timiris Mound Complex, the preferable location of this species off Mauritania, where it achieves a larger size at the ridge, forming the unique very dense assemblages known for the region (**Figures 3** and **6**). Timiris Mound Complex is part of the world longest coral mound barrier, located parallel to the slope at ~500 m depth. Small abundant outcrops of the barrier, which used to be a deep-water coral reef, are present at Timiris. This barrier is providing potential settlement ground for the proliferation of this species at the upper bathyal [11, 29]. Local favorable factors to the growth of this octocoral may promote the monospecific coral garden predominance, despite the dormant state of the coral mounds [32]. Coral mounds have higher framework composed of *Desmophyllum pertusum* (Linnaeus, 1758) [100], increasing the surface for settling of *S. phaeton* larvae, and its access to food in the water column. Suspension feeders on mounds are favored by their location on the limit of water masses, where their food accumulates after transport by tidal currents and downwelling internal waves [53, 54]. High quantity of food at coral mound depths might compensate the presence of an oxygen minimum zone (OMZ) at the same depth. Moreover, the thermocline and the halocline located at ~450 m depth [11] are the upper limit to coral gardens of *S. phaeton* found at coral mounds.

Other extensive coral gardens of *S. phaeton* were discovered at the Nouamghar Canyon, where advantages for the settlement and development of corals, such as hard substrate and current-exposed conditions, are present (**Figures 3** and **4**) [30]. Deep submarine canyons are affected by dense shelf water cascading, which transport organic matter and oxygen-rich waters from the surface to deeper waters, supplying deep-sea filter and suspension feeders with higher quantities of nutritive and fresh plankton, favoring their development [32, 55]. Mauritanian canyons harbor scleractinians at least since the Late Holocene and might be acting as guardians of colonies that will be the source of larvae for the recolonization of neighbor coral mounds and also refuge from unfavorable conditions at the Mauritanian slope.

Geomorphological distribution patterns found herein might be dependent on location with canyon and canyon flank meaning instead Nouamghar, and coral mound and ridge between mounds meaning Timiris (**Figure 4D**). Despite the diverse geomorphology found at the Mauritanian slope, small and medium *S. phaeton* colonies occur preferably at canyon flanks and larger colonies on ridges between coral mounds (**Figure 6**). Exposed flanks and ridges are topographic elevations, essential for settlement of corals, with high food supply by local accelerated hydrography [56, 57]. Still, distinct sizes of *S. phaeton* might be related to 1) long-lasting lack of new recruits of *S. phaeton* on coral mounds or 2) local coral communities and their competition with *S. phaeton*. No data on the growth rate of *Swiftia* species are available, but plexaurid gorgonians have slow growth rates [58], meaning that larger colonies of *S. phaeton* present on mounds can be much older than the younger small/medium colonies on canyons. Probably, for a long time there has been no recruitment or settlement of *S. phaeton* on mounds. On the other side, Nouamghar Canyon has abundant live colonies of reef-forming scleractinians [30] and other predominant undescribed Plexauridae species, which form multispecific coral gardens with *S. phaeton*. However, the Timiris mounds, by being related to a younger aggradation period of coral mound [32], have less scleractinian occurrence. Therefore, there might be less competition with *S. phaeton*, less sediment and consequent impact on the gorgonian survival. Both factors could contribute to the presence of larger colonies at Timiris Mound Complex. Other factors might also be influencing distinct distribution of sizes of this species such as unfavorable conditions for larvae, grazing pressure, and spawning events after fishing impact or even environmental changes.

Coral gardens of *S. phaeton* are also associated with framework formed by *Desmophyllum pertusum* (Linnaeus, 1758) [100] and *Madrepora oculata* Linnaeus, 1758 [100]. Considering that most of the video annotations were made on hard substrate, the results found are potentially biasing (**Figure 6**). Anyway, hard substrate (coral framework, coral rubble, and rocky substrate) is essential for gorgonians settlement and for development of coral gardens, as it is for anemones and sponges [11, 59]. The high species richness of the tropics is also associated with hard substrate on coral reefs [60]. However, *S. phaeton* does not seem to be able to grow to larger sizes if framework is its substrate (**Figure 6**). This might reveal 1) lack of stability of high framework, an irregularly shaped substrate, which is needed for growth 2) competition in dominance between coral reef and coral garden-forming species. If *D. pertusum* forms live dense assemblages, *S. phaeton* is not able to settle its larvae, grow to larger sizes, and has the same access to food. Yet, if coral rubble and rock are the substrate, they are the base for settlement of the gorgonian larvae and are stable for the development of its holdfast and the growth of larger colonies.

In summary, coral gardens of *S. phaeton* are preferably formed on mound ridges and canyon flanks as well as on coral rubble and framework (**Figure 6**). Large colonies forming coral gardens were found mostly on ridges and coral rubble, which are favorable and undisturbed locations with food availability during their life time. Future expeditions looking for coral gardens can now consider these types of locations and geomorphologies as targets of exploration.

During MSM 16/3—Phaeton, the oceanographic data collected were not sufficient to analyze its influence on the distribution of coral gardens formed by *S. phaeton* at a local level [30]. Nevertheless, oceanography explains biogeographical patterns on the distribution of corals. Scleractinians in the NE Atlantic Ocean are influenced by the Equatorial Countercurrent [61], and the structure of primnoid octocorals of Antarctica is related to a natural "soft" biogeographic barrier, the Antarctic Circumpolar Current [62]. In Mauritania, the upper bathyal (400–600 m depth) is influenced by the water masses NACW, north of the Cape Verde Frontal Zone, and SACW, south of the Cape Verde Frontal Zone [48]. Below 300 m depth is the lower SACW characterized by more nutrients and less oxygen, forming the OMZ until 700 m depth [32, 49]. However, at 450 m depth, saline water sits on top of the AAIW forming a layer with a thermocline and a halocline that limits the coral gardens of *S. phaeton* on coral mounds of Mauritania. There, these assemblages thrive at lower temperatures and salinities and are found shallower (480–560 m depth) than in canyons (520–640 m depth). In conjunction with a pronounced seasonal nepheloid layer, both thermocline and halocline promote the development of suspension feeders with high quantity of particulate organic matter [32, 49, 63].

Mauritania is a major upwelling region with a post-glacial oceanography less appropriate for reef-forming scleractinian development, in comparison with the oceanography during glacial times, or at the current northern latitudes [29, 64].

Hydroclimatic condition complexity and paleogeography off Mauritania explain the absence of more diversity and abundance of corals in the deep sea [60], since corals need not only high productivity at the ocean surface, but also lower temperature and higher oxygen concentration for a persistent growth [54]. Climate seems to be related to the distribution of Plexauridae species in the NE Atlantic Ocean, as indicated by distinct species found at temperate (most Macaronesian archipelagos) and tropical waters (Cape Verde) [36]. This is a zoogeographical pattern revealed by several other marine taxa of this area [65]. At the subtropical Mauritania, the Plexauridae are the dominant family of octocorals, but it has distinct diversity than the deep sea of the central NE Atlantic and the Cape Verde Archipelago [36, 37]. Moreover, a nonexhaustive taxonomic assessment of few samples of Plexauridae from Cape Verde and Mauritania revealed different fauna between regions (Sampaio personal observation). The North West African Upwelling acts as a biogeographic barrier for species dispersal between the mainland of Africa and Cape Verde [65]. Also, below the Cape Verde Frontal Zone (CVFZ), where the SACW is located [32], faunistic changes occur and the coral gardens of *S. phaeton* stand in Mauritania at the northern alternance region of the tropical zone, which is characterized by strong seasonal contrasts [60]. In any case, before drawing a portrayal of biogeographical patterns for these regions, higher sampling at unexplored areas and taxonomic revisions are of crucial need to understand the Octocorallia of the deep NW African region and help to predict spatial range changes of deep-sea species under future climate change scenarios.

### 4.1 Natural and anthropogenic impacts on coral gardens

The oldest collection of *S. phaeton* off Mauritania dates back to 1988. After filming *S. phaeton* abundance during *Phaeton* expedition in 2010, specimens deposited at Naturalis, caught during Tyro Mauritania II expedition, were discovered during a visit to the museum [36]. The lack of knowledge on past abundance of this species hampers conclusions on the real natural and anthropogenic impacts on the ecosystems it forms. Hence, we might still infer few impacts affecting these coral gardens. Despite depleted oxygen supply, which can prompt local extinction of benthic fauna, *S. phaeton* is thriving in Mauritania. Yet, ongoing climate change can threaten this ecosystem with further increased oceanic deoxygenation [66]. Sedimentation caused by natural landslides and mechanic anthropogenic events, as the local oil exploration and demersal fisheries, can interfere with the survival of corals, which are suspension feeders. Eolian sediments are deposited off Mauritania by numerous landslides associated with the canyons, especially offshore the Sahara Desert [32, 67]. Known effects caused by sedimentation and oil exposure on corals are the death of the colonies, the shift of its feeding behavior, growth and reproduction, or the disruption on the movement of the polyps and on the calcification of the coral [68, 69]. Sedimentation plumes formed by hydrocarbons-related explorations and bottom trawling are not stagnant and may travel downward, thus affecting deeper communities [70]. *Swiftia exserta* (Ellis & Solander, 1786) [93] forms coral gardens at the Caribbean Sea [71], and it was impacted by the Deepwater Horizon oil spill of the Gulf of Mexico [72–74]. Almost half of the large colonies were injured: broken, with bare branches, overgrowth by hydroids, or covered by sediments below the oil spill [76–78]. Feeding on oil-derived marine snow that sank to the bottom with the pelagic food was also disruptive of the octocoral survival [75]. Gulf of Mexico and Mauritania have hydrocarbon fields [11, 76]. Off Mauritania, the most extensive area of coral gardens of *S. phaeton* (Nouamghar canyon and Timiris Mound Complex) is located on the Block 6 at the area designed for hydrocarbon exploration [11]. The

Chinguetti oil field is situated further to the south (Block 4) and deeper at 800 m depth; however, sediment plumes of the extraction might have traveled toward the coral gardens. Recurrent trawling might increase sedimentation over coral mounds too, reducing settlement, development, and survival of sessile epifauna [29]. When recurrent, trawling reduces the time for continuous growth of colonies and hinders recolonization by new colonies at trawled areas [77]. Longline and trawling fisheries of fish and crustaceans occur in the area since the 1960s, exploring the upper bathyal, where *S. phaeton* lives. Moreover, trawling occurs near the Chinguetti oil field around 600 m depth. On fisheries targeting the Pink spiny lobster *Palinurus mauritanicus* Gruvel, 1911 [97], by-catch of scleractinians was recorded [11]. Yet, fishermen avoid reefs to protect their fishing gears, and while fisheries might have modified the seabed in the past, track marks are far from mound structures [29].

### 4.2 Definition of octocoral garden for conservation

The current definition of coral garden used by the Oslo-Paris Convention (OSPAR) in alignment with the 61/105 [16] aiming to protect deep-sea VMEs from anthropogenic pressures includes all the coral taxa, except the reef-building scleractinians [78]. Nevertheless, the term "coral garden" was initially applied specifically to dense assemblages of octocorals in cold waters [9]. Increasing the number of taxa within this definition led to a higher complexity of assemblages, biotopes, and substrates caught within the coral garden concept. When the density of these assemblages is not considered to establish boundaries on their distribution, it is difficult to clearly isolate the idea of what is a coral garden and apply conservation measures [79]. A density criterion of >0.1 colonies $m^{-2}$ across an area of at least 25 $m^2$ was added to the coral garden definition: areas where the coral garden-forming species forms dense aggregations [19, 79, 80]. ICES [81] had previously tried to include the density of corals to remove arbitrary decisions on this term, but it was difficult to quantify densities of corals *in situ*. This is commonly the case with deep-sea exploratory cruises, not aiming to quantify assemblages, but to understand what taxonomic groups inhabit the area. On the other hand, it is hard to calculate the area observed, without considering several measurements while annotating ROV images [82]. In the absence of a calculated area on the ROV images from Mauritania, we used a proxy of the species density: classes of relative abundance related to its size. Coral gardens were then defined when dense and very dense assemblages were annotated. Herein, we added the size to the density on the concept of coral garden, because the size of a species is important to define the scale of the ecosystem it forms. If a gorgonian has a smaller maximum size, as it is the case of *S. phaeton*, it will consequently form coral gardens on a smaller scale and area, than a species with a higher maximum size. Therefore, instead of stipulating 25 $m^2$ for all the species, the definition of coral garden should indeed consider areas, which take into account the maximum size a species can achieve. The long-lasting discussion on the definition of terrestrial forests should be applied to biotope concepts in marine sciences. Current concepts of forest vary, but already include the size of the species and the size of the forests it forms [83, 84].

## 5. Conclusions

The ongoing effort to define marine-protected areas (MPAs) off Mauritania should now consider the locations were *Swiftia phaeton* forms coral gardens in higher density and longer areas. The Timiris Mound Complex, where the species thrives, is particularly important, and it is located nearby an oil exploration field. Therefore, Timiris should be protected from hydrocarbon exploration and trawling fisheries. Future cruises and studies should visit it along with other locations where the species is less abundant, in order to collect local oceanographic data and infer the influence of currents and climate, on the distribution of this species and its coral gardens.

Coral gardens of *S. phaeton* harbor several kinds of taxa that may depend on this species to feed, hide, or reproduce. This flag species of gorgonian, commonly associated with *D. pertusum*, should be a priority for conservation in Mauritania, where it forms the most abundant assemblages of the genus *Swiftia*, unique at the NE Atlantic and Mediterranean scale.

## Acknowledgements

Authors are thankful to the crew and scientific team of MSM 16/3—Phaeton cruise onboard RV *Maria S. Merian* for collecting images and data that made this work possible. Special thanks are devoted to the ROV Pilot Thomas Lundälv from the Sven Lovén Center for Marine Infrastructure at Tjarnö, Sweden, for his skillful work and the copyright of underwater images (**Figures 2** and 7). Thanks to Maria Rakka for statistical guidelines, Carlos Vila-Viçosa for the botanical literature, and to the referees for improving the manuscript.

ÍS was funded by Fundação para a Ciência e a Tecnologia (FCT) Doctoral grant SFRH/BD/101113/2014. This research received support from the SYNTHESYS Project http://www.synthesys.info/ which is financed by European Community Research Infrastructure Action under the FP7 "Capacities" Program. This publication is supported by the Deutsche Gesellschaft für Internationale Zusammenarbeit (GIZ) through the WASP (West African Biodiversity under Pressure) Project, Contract 81248171, to AF.

## Author details

Íris Sampaio[1,2]*, Lydia Beuck[1], Gui M. Menezes[3] and André Freiwald[1,4]

1 Marine Research Department, Senckenberg am Meer, Wilhelmshaven, Germany

2 University of the Azores, Horta, Portugal

3 Institute of Marine Sciences—Okeanos, University of the Azores, Horta, Portugal

4 MARUM—Center for Marine Environmental Sciences, Bremen, Germany

*Address all correspondence to: irisfs@gmail.com

## References

[1] Beaman RJ, Daniell JJ, Harris PT. Geology–benthos relationships on a temperate rocky bank, eastern Bass Strait, Australia. Marine and Freshwater Research. 2005;**56**(7):943-958. DOI: 10.1071/MF04306

[2] Heifetz J, Wing BL, Stone RP, Malecha PW, Courtney DL. Corals of the Aleutian islands. Fisheries Oceanography. 2005;**14**:131-138

[3] George RY, Okey TA, Reed JK, Stone RP. Ecosystem-based fisheries management of seamount and deep-sea coral reefs in US waters: Conceptual models for proactive decisions. Bulletin of Marine Science. 2007;**81**(Supplement 1):9-30

[4] Long DJ, Baco AR. Rapid change with depth in megabenthic structure-forming communities of the Makapu'u deep-sea coral bed. Deep-Sea Research Part II. 2014;**99**:158-168

[5] Tempera F, Pereira JN, Henriques AB, Porteiro F, Morato T, Matos V, et al. Cataloguing deep-sea biological facies of the Azores. Revista de Investigaciones Marinas. 2012;**19**:36-38

[6] Sheehan EV, Rees A, Bridger D, Williams T, Hall-Spencer JM. Strandings of NE Atlantic gorgonians. Biological Conservation. 2017;**209**:482-487

[7] Morato T, Dominguez-Carrió C, Mohn C, Vicente OO, Ramos M, Rodrigues L, et al. Dense cold-water coral garden of *Paragorgia johnsoni* suggests the importance of the Mid-Atlantic Ridge for deep sea biodiversity. Ecology and Evolution, Nature Notes. 2021;**23**:16426-16433

[8] Roberts JM, Wheeler AJ, Freiwald A, Cairns SD. Cold-Water Corals. Cambridge: University Press; 2009

[9] Auster PJ, Kilgour M, Packer D, Waller R, Auscavitch S, Watling L. Octocoral gardens in the Gulf of Maine (NW Atlantic). Biodiversity. 2013;**14**(4):193-194

[10] Elverhøi A, Norem H, Andersen ES, Dowdeswell JA, Fossen I, Haflidason H, et al. On the origin and flow behavior of submarine slides on deep-sea fans along the Norwegian–Barents Sea continental margin. Geo-Marine Letters. 1997;**17**(2):119-125. DOI: 10.1007/s003670050016

[11] Colman JG, Gordon DM, Lane AP, Forde MJ, Fitzpatrick JJ. Carbonate mounds off Mauritania, Northwest Africa: Status of deep-water corals and implications for management of fishing and oil exploration activities. In: Freiwald A, Roberts MJ, editors. Cold-Water Corals and Ecosystems. Springer, Berlin: Heidelberg; 2005. pp. 417-441. DOI: 10.1007/3-540-27673-4_21

[12] Althaus F, Williams A, Schlacher TA, Kloser RJ, Green MA, Barker BA, et al. Impacts of bottom trawling on deep-coral ecosystems of seamounts are long-lasting. Marine Ecology Progress Series. 2009;**397**:279-294. DOI: 10.3354/meps08248

[13] Van Dover CL, Ardron JA, Escobar E, Gianni M, Gjerde KM, Jaeckel A. Biodiversity loss from deep-sea mining. Nature Geoscience. 2017;**10**:464

[14] Freiwald A, Fosså JH, Grehan A, Koslow T, Murray JR. Cold-Water Coral Reefs: Out of Sight No Longer Out of Mind. Cambridge, UK: UNEP-WCMC; 2004

[15] United Nations General Assembly. Oceans and the Law of the Sea. Report

of the Secretary General; A/58/65:26668. 2003

[16] UNGA. Resolution 61/105 sustainable fisheries, including through the 1995 agreement for the implementation of the provisions of the united nations convention on the law of the Sea of 10 December 1982 relating to the conservation and management of straddling fish stocks and highly migratory fish stocks, and related instruments. Operative Paragraphs. 2006. pp. 80-91

[17] Fabri MC, Pedel L, Beuck L, Galgani F, Hebbeln D, Freiwald A. Megafauna of vulnerable marine ecosystems in French Mediterranean submarine canyons: Spatial distribution and anthropogenic impacts. Deep-Sea Research Part II. 2014;**104**:184-207. DOI: 10.1016/j.dsr2.2013.06.016

[18] FAO. International Guidelines for the Management of Deep-Sea Fisheries in the High Seas. Roma; 2009. Available from: http://www.fao.org/3/i0816t/i0816t00.htm [Accessed: 09 September 2019]

[19] OSPAR. Background Document for Coral Gardens. Biodiversity Series. 2010

[20] Ramirez-Llodra E, Tyler PA, Baker MC, Bergstad OA, Clark MR, Escobar E, et al. Man and the last great wilderness: Human impact on the deep sea. PLoS One. 2011;**6**(8):e22588. DOI: 10.1371/journal.pone.0022588

[21] Galparsoro I, Connor DW, Borja A, Aish A, Amorim P, Bajjouk T, et al. Using EUNIS habitat classification for benthic mapping in European seas: Present concerns and future needs. Marine Pollution Bulletin. 2012;**64**(12):2630-2638. DOI: 10.1016/j.marpolbul.2012.10.010

[22] Davies JS, Guillaumont B, Tempera F, Vertino A, Beuck L, Ólafsdóttir SH, et al. A new classification scheme of European cold-water coral habitats: Implications for ecosystem-based management of the deep sea. Deep-Sea Research Part II. 2017;**145**:102-109. DOI: 10.1016/j.dsr2.2017.04.014

[23] Bigelow HB. The work of the 'Michael Sars' in the North Atlantic in 1910. Science. 1911;**34**(862):7-10

[24] Den Hartog JC. An introduction to the CANCAP-project of the Dutch Rijksmuseum von natuurlijke historie (RMNH), with special reference to the CANCAP-VI expedition (1982) to the Cape Verde islands. Courier Forschungsinstitut Senckenberg. 1984;**68**:515

[25] Van der Land J. *Tyro* Mauritania-II Expedition List of Stations: 1-9. Leiden: Rijksmuseum an Natuurlijke Historie; 1988 [mimeographed]

[26] Oñate SC. Marine molluscs (Gastropoda and Bivalvia) from Northwest Africa [PhD thesis]. Vigo: Universidade de Vigo; 2017

[27] Ramos A, Ramil F, Sanz JL. Deep-sea ecosystems off Mauritania: An introduction. In: Ramos A, Ramil F, Sanz JL, editors. Deep-Sea Ecosystems Off Mauritania. Dordrecht: Springer; 2017. pp. 1-51

[28] Ramiro-Sánchez B, González-Irusta JM, Henry LA, Cleland J, Yeo I, Xavier JR, et al. Characterization and mapping of a deep-sea sponge ground on the Tropic Seamount (northeast tropical Atlantic): Implications for spatial management in the high seas. Frontiers in Marine Science. 2019;**6**:278

[29] Ramos A, Sanz JL, Ramil F, Agudo LM, Presas-Navarro C. The giant cold-water coral mounds barrier off Mauritania. In: Ramos A, Ramil F, Sanz JL, editors. Deep-Sea Ecosystems

Off Mauritania. Dordrecht: Springer; 2017. pp. 481-525

[30] Westphal H, Beuck L, Braun S, Freiwald A, Hanebuth T, Hetzinger S, Klicpera A, Kudrass H, Lantzsch H, Lundälv T, Vicens GM, Preto N, Reumont J, Schilling S, Taviani M, Wienberg C. Phaeton—Paleoceanographic and Paleo-Climatic Record on the Mauritanian Shelf. Cruise No. MSM16/3 in Maria S. Merian-Berichte. Hamburg: Leitstelle Deutsche Forschungsschiffe Institut für Meereskunde der Universität Hamburg). DFG-Senatskommission für Ozeanographie; 2012

[31] Ramos A, Ramil F, Sanz JL. Deep-Sea Ecosystems Off Mauritania. Research of Marine Biodiversity and Habitats in the Northwest African Margin. The Netherlands: Springer; 2017

[32] Wienberg C, Titschack J, Freiwald A, Frank N, Lundälv T, Taviani M, et al. The giant Mauritanian cold-water coral mound province: Oxygen control on coral mound formation. Quaternary Science Reviews. 2018;**185**:135-152

[33] Le Danois E. Les profondeurs de la mer: trente ans de recherches sur la faune sous-marine au large des côtes de France. Paris: Payot; 1948

[34] Buhl-Mortensen P, Buhl-Mortensen L. Diverse and vulnerable deep-water biotopes in the Hardangerfjord. Marine Biology Research. 2014;**10**(3):253-267. DOI: 10.1080/17451000.2013.810759

[35] Gori A, Bavestrello G, Grinyó J, Dominguez-Carrió C, Ambroso S, Bo M. Animal forests in deep coastal bottoms and continental shelves of the Mediterranean sea. In: Rossi S, Bramanti L, Gori A, Orejas C, editors. Marine Animal Forests: The Ecology of Benthic Biodiversity Hotspots. Cham, Switzerland: Springer; 2017. pp. 207-213

[36] Sampaio Í, Carreiro-Silva M, Freiwald A, Menezes G, Grasshoff M. Natural history collections as a basis for sound biodiversity assessments: Plexauridae (Octocorallia, Holaxonia) of the Naturalis CANCAP and Tyro Mauritania II expeditions. ZooKeys. 2019;**870**:1. DOI: 10.3897/zookeys.870.35285

[37] Grasshoff M. Die Gorgonarien des östlichen Nordatlantik und des Mittelmeeres: III. Die Familie Paramuriceidae (Cnidaria: Anthozoa). Meteor – Forschungs – Ergebnisse. 1977;**D27**:5-76

[38] Sampaio Í, Beuck L, Freiwald A. A new octocoral species of *Swiftia* (Holaxonia, Plexauridae) from the upper bathyal off Mauritania (NE Atlantic). ZooKeys. 2022;**1106**:121-140. DOI: 10.3897/zookeys.1106.81364

[39] Sanz JL, Maestro A, Agudo LM. The Mauritanian margin. Bathymetric and geomorphological characteristics. In: Ramos A, Ramil F, Sanz JL, editors. Deep-Sea Ecosystems Off Mauritania. Dordrecht: Springer; 2017. pp. 53-117. DOI: 10.1007/978-94-024-1023-5_2

[40] Matsuyama K, Titschack J, Baum D, Freiwald A. Two new species of erect Bryozoa (Gymnolaemata: Cheilostomata) and the application of non-destructive imaging methods for quantitative taxonomy. Zootaxa. 2015;**4020**:81-100

[41] Gil M, Ramil F, Ansín AJ. Hydroids (Cnidaria, Hydrozoa) from Mauritanian coral mounds. Zootaxa. 2020;**4878**:412-466

[42] Göcke C, Hestetun JT, Uhlir C, Freiwald A, Beuck L, Janussen D. *Cladorhiza corallophila* sp. nov., a new carnivorous sponge (Cladorhizidae,

Demospongiae) living in close association with *Lophelia pertusa* and *Madrepora oculata* (Scleractinia). Zootaxa. 2016;**4168**:512-524

[43] De Matos-Pita SS, Ramil F, Ramos A. Marine lobsters and lithodids (Crustacea: Decapoda) from Mauritanian deep-waters (NW Africa). Regional Studies in Marine Science. 2018;**23**:32-38

[44] Hoffman L, Freiwald A. *Cantrainea mauritanea* n. sp. (Gastropoda: Colloniidae) from deep-water coral habitats off Mauritania. Miscellanea Malacologica. 2020;**8**:69-73

[45] Frank N, Freiwald A, Correa ML, Wienberg C, Eisele M, Hebbeln D, et al. Northeastern Atlantic cold-water coral reefs and climate. Geology. 2011;**39**(8):743-746

[46] Messié M, Chavez FP. Seasonal regulation of primary production in eastern boundary upwelling systems. Progress in Oceanography. 2015;**134**:1-18

[47] Stramma L, Hüttl S, Schafstall J. Water masses and currents in the upper tropical northeast Atlantic off Northwest Africa. Journal of Geophysical Research, Oceans. 2005;**110**:C12006

[48] Pastor MV, Peña-Izquierdo J, Pelegrí JL, Marrero-DÍaz A. Meridional changes in water mass distributions off NW Africa during November 2007/2008. Ciencias Marinas. 2012;**38**(1B):223-244

[49] Pelegrí JL, Peña-Izquierdo J, Machín F, Meiners C, Presas-Navarro C. Oceanography of the Cape Verde Basin and Mauritanian Slope Waters. In: Ramos A, Ramil F, Sanz JL, editors. Deep-Sea Ecosystems Off Mauritania. Dordrecht: Springer; 2017. pp. 119-153. DOI: 10.1007/978-94-024-1023-5_3

[50] Stramma L, Brandt P, Schafstall J, Schott F, Fischer J, Körtzinger A. Oxygen minimum zone in the North Atlantic south and east of the Cape Verde islands. Journal of Geophysical Research, Oceans. 2008;**113**:C4

[51] Wintersteller P, Titschack J, Gaide S, Hanebuth TJJ, Freiwald A, Westphal H. Gridded EM120 Multibeam-Echosounder Bathymetry of Cruise MSM16-3. Pangaea: MARUM-Center for Marine Environmental Sciences, University Bremen; 2017

[52] Hoffman L, Beuck L, Van Heugten B, Lavaleye M, Freiwald A. Last snails standing since the Pleistocene, a tale of Calliostomatidae (Gastropoda) living in deep-water coral habitats in the northeast Atlantic. Zootaxa. 2019;**4613**:93-110

[53] Henry LA, Roberts JM. Biodiversity and ecological composition of macrobenthos on cold-water coral mounds and adjacent off-mound habitat in the bathyal Porcupine Seabight. NE Atlantic. Deep-Sea Research PT I. 2007;**54**(4):654-672

[54] Eisele M, Frank N, Wienberg C, Hebbeln D, Correa ML, Douville E, et al. Productivity controlled cold-water coral growth periods during the last glacial off Mauritania. Marine Geology. 2011;**280**(1-4):143-149

[55] Canals M, Puig P, de Madron XD, Heussner S, Palanques A, Fabres J. Flushing submarine canyons. Nature. 2006;**444**(7117):354

[56] Roberts JM, Wheeler AJ, Freiwald A. Reefs of the deep: The biology and geology of cold-water coral ecosystems. Science. 2006;**312**:543-547

[57] Tamborrino L, Wienberg C, Titschack J, Wintersteller P, Mienis F, Schröder-Ritzrau A, et al. Mid-Holocene extinction of cold-water corals on the Namibian shelf steered by the Benguela

oxygen minimum zone. Geology. 2019;**47**:14

[58] Prouty NG, Fisher CR, Demopoulos AW, Druffel ER. Growth rates and ages of deep-sea corals impacted by the deepwater horizon oil spill. Deep-Sea Research PT II. 2016;**129**:196-212

[59] Roberts JM, Henry LA, Long D, Hartley JP. Cold-water coral reef frameworks, megafaunal communities and evidence for coral carbonate mounds on the Hatton bank, north east Atlantic. Facies. 2008;**54**(3):297-316

[60] Le Lœuff P, von Cosel R. Biodiversity patterns of the marine benthic fauna on the Atlantic coast of tropical Africa in relation to hydroclimatic conditions and paleogeographic events. Acta Oecologica. 1998;**19**(3):309-321

[61] Cairns SD, Chapman RE. Biogeographic affinities of the North Atlantic deep-water Scleractinia. In: Proceedings of the First International Symposium on Deep-sea Corals. Halifax, Nova Scotia: Ecology Action Centre and Nova Scotia Museum; 2001. pp. 30-57

[62] Dueñas LF, Tracey DM, Crawford AJ, Wilke T, Alderslade P, Sánchez JA. The Antarctic circumpolar current as a diversification trigger for deep-sea octocorals. BMC Evolutionary Biology. 2016;**16**(1):2. DOI: 10.1186/s12862-015-0574-z

[63] Fischer G, Reuter C, Karakas G, Nowald N, Wefer G. Offshore advection of particles within the Cape Blanc filament, Mauritania: Results from observational and modelling studies. Progress in Oceanography. 2009;**83**:322e330. DOI: 10.1016/j.pocean.2009.07.023

[64] Freiwald A. Messinian salinity crisis: What happened to cold-water corals? In: Orejas C, Jiménez C, editors. Mediterranean Cold-Water Corals: Past, Present and Future. Coral Reefs of the World. Vol. 9. Cham: Springer; 2019. pp. 47-50

[65] Freitas R, Romeiras M, Silva L, Cordeiro R, Madeira P, González JA, et al. Restructuring of the 'Macaronesia' biogeographic unit: A marine multi-taxon biogeographical approach. Scientific Reports. 2019;**9**(1):1-18

[66] Sweetman AK, Thurber AR, Smith CR, Levin LA, Mora C, Wei CL, et al. Major impacts of climate change on deep-sea benthic ecosystems. Elementa Science of the Anthropocene. 2017;**5**:4

[67] Harrison SP, Kohfeld KE, Roelandt C, Claquin T. The role of dust in climate changes today, at the last glacial maximum and in the future. Earth-Science Reviews. 2001;**54**:43e80

[68] Rogers AD. The biology of *Lophelia pertusa* (Linnaeus 1758) and other deep-water reef-forming corals and impacts from human activities. International Review of Hydrobiology. 1999;**84**(4):315-406

[69] Goodbody-Gringley G, Wetzel DL, Gillon D, Pulster E, Miller A, Ritchie KB. Toxicity of deepwater horizon source oil and the chemical dispersant, Corexit® 9500, to coral larvae. PLoS One. 2013;**8**(1):e45574. DOI: 10.1371/journal.pone.0045574

[70] Martín J, Puig P, Palanques A, Ribó M. Trawling-induced daily sediment resuspension in the flank of a Mediterranean submarine canyon. Deep-Sea Research Part II. 2014;**104**:174-183

[71] Hourigan TF, Lumsden SE, Dorr G, Bruckner AW, Brooke S, Stone RP. State of deep coral ecosystems of the United States: Introduction and national overview. The State of deep

coral ecosystems of the United States. Silver Spring, MD: NOAA Technical Memorandum CRCP-3; 2007. pp. 1-64

[72] Etnoyer PJ, Wickes LN, Silva M, Dubick JD, Balthis L, Salgado E, et al. Decline in condition of gorgonian octocorals on mesophotic reefs in the northern Gulf of Mexico: Before and after the deepwater horizon oil spill. Coral Reefs. 2016;**35**(1):77-90. DOI: 10.1007/s00338-015-1363-2

[73] Frometa J, DeLorenzo ME, Pisarski EC, Etnoyer PJ. Toxicity of oil and dispersant on the deep water gorgonian octocoral *Swiftia exserta*, with implications for the effects of the deepwater horizon oil spill. Marine Pollution Bulletin. 2017;**122**(1-2):91-99. DOI: 10.1016/j.marpolbul.2017.06.009

[74] Silva M, Etnoyer PJ, MacDonald IR. Coral injuries observed at mesophotic reefs after the deepwater horizon oil discharge. Deep-Sea Research Part II. 2016;**129**:96-107

[75] Passow U. Formation of rapidly-sinking, oil-associated marine snow. Deep-Sea Research Part II. 2016;**129**:232-240

[76] Hovland M, Mortensen PB, Brattegard T, Strass P, Rokoengen K. Ahermatypic coral banks off mid-Norway: Evidence for a link with seepage of light hydrocarbons. PALAIOS. 1998;**13**:189-200. DOI: 10.2307/3515489

[77] Fosså JH, Mortensen PB, Furevik DM. The deep-water coral *Lophelia pertusa* in Norwegian waters: Distribution and fishery impacts. Hydrobiologia. 2002;**471**(1-3):1-12

[78] OSPAR. Descriptions of habitats on the OSPAR list of threatened and/or declining species and habitats. OSPAR Agreement 2008-07. 2008

[79] Bullimore RD, Foster NL, Howell KL. Coral-characterized benthic assemblages of the deep Northeast Atlantic: Defining "coral gardens" to support future habitat mapping efforts. ICES Journal of Marine Science. 2013;**70**(3):511-522

[80] Rogers AD, Kemp K, Davies AJ, Taylor ML. The diseases of deep-water corals. In: Woodley CM, Downs CA, Bruckner AW, Porter JW, Galloway SB, editors. Diseases of Corals. London, New York: CRC Press; 2013. pp. 416-441

[81] ICES. Report of the Working Group on Deep-Water Ecology (WGDEC). 26-28 February 2007. ICES Advisory Committee on Ecosystems. ICES Document CM 2007/ACE: 01; 2007

[82] Dias FC, Gomes-Pereira J, Tojeira I, Souto M, Afonso A, Calado A, et al. Area estimation of deep-sea surfaces from oblique still images. PLoS One. 2015;**10**(7):e0133290. DOI: 10.1371/journal.pone.0133290

[83] Rivas-Martínez S, Díaz TE, Fernández-González F, Izco J, Loidi J, Lousã M, et al. Vascular plant communities of Spain and Portugal. Addenda to the Syntaxonomical checklist of 2001. Itinera Geobotanica. 2002;**15**(1-2):5-922

[84] Chazdon RL, Brancalion PHS, Laestadius L, Bennett-Curry A, Buckingham K, Kumar C, et al. When is a forest a forest? Forest concepts and definitions in the era of forest and landscape restoration. Ambio. 2016;**45**(5):538-550

[85] ESRI Ocean Basemap. ESRI Ocean Basemap. 2019. Available from: https://www.arcgis.com/home/item.html?id=6348e67824504fc9a62976434bf0d8d5

[86] GEBCO Compilation Group. GEBCO 2019 Grid. 2019. Available from: https://doi.org/10.5285/836f016a-33be-6ddc-e053-6c86abc0788e

## Taxonomic authorities

[87] Broch H. Alcyonarien. In: Results of Dr. E. Mjobergs Swedish Scientific Expeditions to Australia 1910-1913. XI. Kgl ed. Handlingar: Svenska Vetenskapsakad; 1916. pp. 1-48

[88] de Saint-Laurent M, Macpherson E. Les espèces atlantiques du genre *Eumunida* Smith, 1883 (Crustacea: Decapoda: Chirostylidae). Journal of Natural History. 1990;**24**:647-666

[89] Delaroche F-É. Suite du mémoire sur les espèces de poissons observées à Iviça. Tableau des espèces de poissons que j'ai observées à Iviça pendant les mois de décembre, janvier et février. Tableau des poissons que j'ai observés à Maïorque et à Barcelonne, mais que je n'ai point vus à Iviça. Observations sur quelques-uns des poissons indiqués dans le précédent tableau, et descriptions des espèces nouvelles ou peu connues. Annales du Muséum d'Histoire Naturelle Paris. 1809;**13**:313-361

[90] Delle Chiaje S. Memorie sulla storia e notomia degli animali senza vertebre del regno di Napoli. Napoli: Società Tipografica; 1830. p. 116

[91] Duchassaing P, Michelotti J. Supplément au mèmoire sur les coralliaires des Antilles. Turin: Mémoires de l'Academie des Sciences de Turin; 1864

[92] Duncan PM. A description of the Madreporaria dredged up during the Expeditions of H.M.S. 'Porcupine' in 1869 and 1870. Transactions of the Zoological Society of London. 1873;**8**:303-344

[93] Ellis J, Solander D. The Natural History of Many Curious and Uncommon Zoophytes Collected from Various Parts of the Globe. Horace's Head, Fleet-Street, London: Benjamin White & Son; 1786

[94] Esper EJC, editor. Die Pflanzenthiere in Abbildungen nach der Natur mit farben erleuchtet nebst Beschreibungen. Raspichen: Nürnberg; 1788-1830

[95] Grasshoff M. Die Gorgonaria der Expeditionen von "Travailleur" 1880-1882 und "Talisman" 1883 (Cnidaria, Anthozoa). Bulletin du Muséum national d'histoire naturelle. 1986;**4**:9-38

[96] Gray JE. On the arrangement of zoophytes with pinnated tentacles. Annals and Magazine of Natural History. 1859;**4**:439-444. DOI: 10.1080/00222935908697159

[97] Gruvel A. Contribution à l'etude générale systématique et économique des Palinuridae. Annales de l'Institut Océanographique. 1911;**3**:1-56

[98] Koren J, Danielssen DC. Nye Alcyonider, Gorgonider og Pennatulider tilhorende Norges Fauna. Bergens Museum Skrifter. 1883;**2**:1-38

[99] Lamarck JD, editor. Histoire naturelle des animaux sans vertèbres. Paris: Verdière; 1816

[100] Linnaeus C. Systema naturae (Editio Decima). Holmiae, Impensis Direct. Stockholm: Laurentii Salvii; 1758

[101] Madsen FJ. Remarks on *Swiftia rosea* (Grieg) and related species. Steenstrupia. 1970;**1**(1):1-10

[102] Risso A, editor. Histoire Naturelle des Crustacés des Environs de Nice. Librairie Grecque-Latine-Allemande: Paris; 1816. p. 175

[103] Thomson JA, Simpson JJ. An Account of the Alcyonarians Collected by the Royal Indian Survey Ship "Investigator" in the Indian Ocean. II. The Alcyonarians of the Littoral Area. Calcutta: The Indian Museum; 1909. p. 319

[104] Thomson JA. Alcyonaires des environs de Monaco et de localités diverses. Bulletin de l'Institut océanographique. 1929;**534**:1-10

Chapter 3

# Oil Spill Incidents on Coral Reefs: Impacts and Remediation Technologies

*Luanny Fernandes, Flávia L. Carmo, Hugo E. de Jesus, Adriana U. Soriano and Henrique F. Santos*

## Abstract

Despite the great ecological and economic importance of coral reefs, these ecosystems are especially sensitive to environmental changes and vulnerable to impacts from various anthropogenic activities, including contamination by oil spills. Oil spills occur worldwide, mostly in marine environments, and have been reported for decades. Furthermore, the main oil transport routes in the oceans are close to important coral reefs and many of the major oil spills in history have occurred near these areas. Because of the widespread use of petroleum products, offshore oil and gas production has significantly increased its potential since the 1990s, thus increasing the risk of accidents in marine environments and consequently on coral reefs. Despite the great risk of oil exploitation to coral reefs, there is still no efficient, sustainable, and large-scale applicable remediation strategy to protect or to clean up reefs impacted by oil spills. Current methodologies to remediate oil pollution in marine environments are based on the use of chemical dispersants; however, these can be more harmful to corals than oil itself. Meanwhile, the use of bioremediation strategy, through the manipulation of the coral microbiome, has been proposed as a possible alternative to mitigate the impacts of oil on coral reefs.

**Keywords:** petroleum, anthropogenic impacts, bioremediation, microbiology, probiotics

## 1. Introduction

Petroleum is a naturally occurring complex mixture composed predominantly of carbon and hydrogen compounds (hydrocarbons) that occur in liquid, gaseous, or solid form. Hydrocarbons form a large variety of complex molecular structures, which can vary in terms of physicochemical properties, toxicity, and biodegradability. Based on their molecular structure, the hydrocarbons can be classified into four main groups: alkanes, alkenes, cycloalkanes, and aromatics [1].

Alkanes, also known as paraffins or saturated aliphatics, are the simplest hydrocarbons and contain only single covalent bonds between carbon atoms.

They comprise the largest fraction of most oils, their toxicity is usually low, and they are easily biodegraded, excepting those that can act as solvents, such as *n*-hexane and *n*-heptane. Alkenes, or olefins, are unsaturated hydrocarbon molecules that contain one or more double bonds between carbon atoms. They are usually in small amounts or absent in oil; however they are highly present in refining products such as gasoline. Cycloalkanes, or naphthenes, are the monocyclic saturated hydrocarbons and they represent the second largest fraction of most oils. Aromatics are cyclic unsaturated hydrocarbons, with one or more aromatic rings in their molecule. When they have more than one aromatic ring, they are called polycyclic aromatic hydrocarbons (PAHs). PAHs contamination is a major environmental concern due to their acute toxic potential, their resistance to biodegradation and their potential to bioaccumulate [1].

Crude oil, the liquid form of petroleum, constitutes the most important primary fossil fuel. It can be found in underground reservoirs, and its accessed and extracted by drilling, on land (onshore reservoirs), or at sea (offshore reservoirs). Once extracted it is refined to produce fuels such as gasoline, diesel, jet and heating fuel, lubricating oils, asphalts and also petrochemical intermediate feedstocks, which are used in the production of a wide range of solvents, plastics, and detergents, among other important products for our modern life. Thus, since the rapid global economic growth has increased the demand for all products mentioned above, oil production, processing, and transportation activities have also experienced an intensification in the last decades, bringing with them higher risk of oil spills, even considering the advance of oil spill prevention technologies.

In addition, with increasing risk of spills occurrence, concerns about the differentiation of oils in terms of composition, environmental behavior, and toxicity have arisen.

In a general manner, it is possible to say that once in the sea, oil begins to suffer a set of transformations, such as evaporation, aerosolization, photooxidation, mixing, emulsification, diffusion, partial water dissolution, spreading, transport, biodegradation, aggregation, adhesion, and sedimentation [2]. The magnitude of those phenomena depends on the physicochemical characteristics, composition, and the amount of oil spilled, as well as on atmospheric and oceanographic conditions, level of particulates and organic matter in suspension, availability of nutrients in the water, and mainly on the type of spill (superficial or subsea). **Figure 1** summarizes most of those phenomena.

The impact of a complex mixture of hydrocarbons on marine organisms affected by a spill event can vary a lot, spatially, among species, and also with time. At the surface, for instance, seabirds, turtles, and marine mammals can be harmed by direct contact, inhalation, and aspiration of oil, instead in the water column, fishes can be impacted and at the seafloor, corals can be strongly damaged [2]. The time variable is also important, since some processes largely occur in the first days or weeks of spill (i.e., photooxidation, evaporation), while others start later and last longer, for months or years (i.e., emulsification, biodegradation). As pollutant composition varies with time and environmental compartments as result of natural processes, it is expected that bioavailability and toxicity also change dynamically. Thus, it is common to find large quantities of light hydrocarbons at the surface and in the water column at the beginning of the spill event ready for evaporation, photooxidation, and even biodegradation, and to find sinking aggregates or seabed deposits of enriched PAH due to incomplete photooxidation in the surface and/or biodegradation in the water column [2].

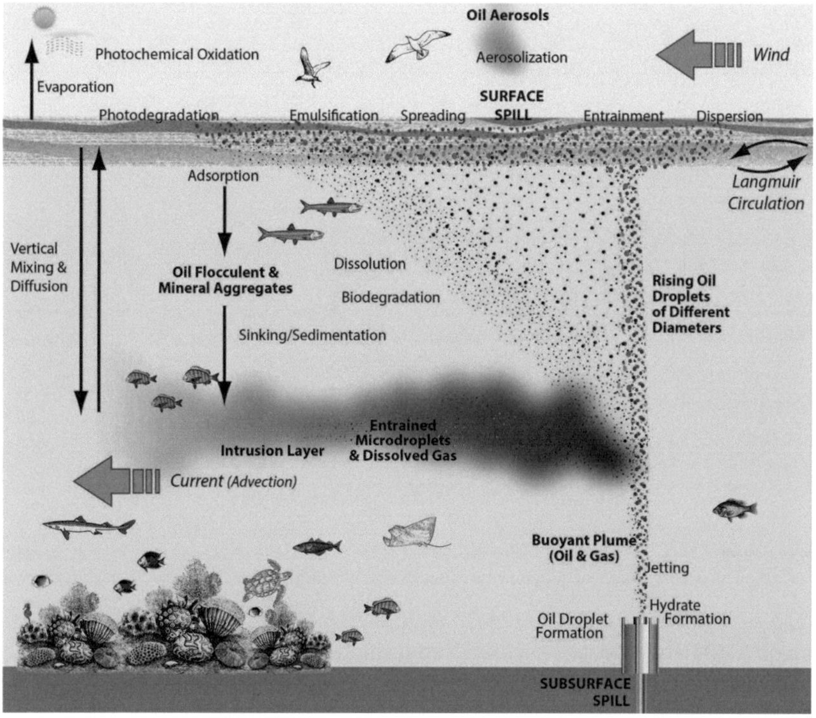


**Figure 1.**
*Summary of phenomena that occur in oil spills and some potential affected marine organisms. Source: Modified from [2].*

## 2. Oil spills in coral reefs

Accidents involving oil or its derivatives can happen during drilling activity, refining, transportation, distribution, or use. Because of the widespread use of petroleum products, oil spills are frequent [3] and have been reported for decades [4–7].

Most oil spills occur in the ocean. The annual amount of oil spilled into the ocean is approximately $10^3$ tons [8]. Unfortunately, the main transport routes for oil in the oceans are close to important coral reefs (**Figure 2**) and many of the major oil spills in history have occurred close to coral reef areas [7].

Besides, due to the worldwide increase in oil and gas demand and the dwindling of onshore reserves, offshore oil, and gas production have significantly increased their potential since the 1990s [9]. This increased the risk of accidents in marine environments and, consequently, in coral reefs.

Until recent days, several incidents involving oil and its derivatives have occurred in coral reef regions, leading to a considerable loss of coral cover.

In April 1986, in Bahia Las Minas (Panamá), a crude oil incident spilled around 10 to 16 million liters of crude oil, causing lethal, and sublethal effects in the coral regions [10]. During the Gulf War, in January 1991, a large spill of 6.3 million barrels also caused irreversible damages to coral reefs in Kuwait and Saudi Arabia [11]. In the Caribbean Sea, a continuing petroleum contamination between 1923 and 1985 by a refinery in Aruba (Venezuela) also resulted in serious damage to corals. Currently, in this region, coral cover is very low, and younger corals are not found within 9 km of the refinery's surroundings [12].

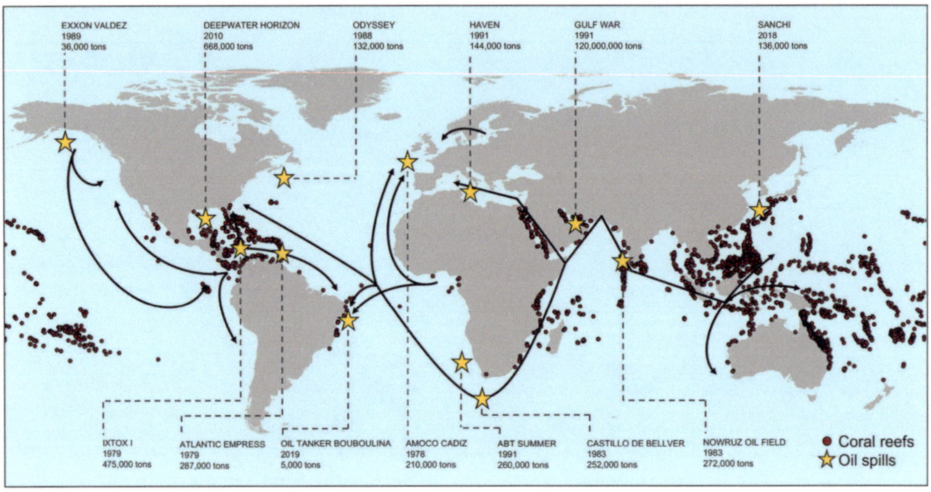


**Figure 2.**
*The figure shows coral reefs on the planet, oil routes in the oceans, and some of the major oil spills. Arrows represent common oil routes, small red circles represent coral reefs, and yellow stars represent oil spills.*

In April 2010, the biggest environmental disaster in the history of the United States, the explosion on the British Petroleum Deepwater Horizon platform, located in the Gulf of Mexico, caused the opening of the oil well, generating a leak of about 1 million liters per day. A study conducted on the impact of this spill on deep-sea corals concluded that the spilled oil devastated corals living about 7 km southwest of the well. The affected coral reef occupies an area equivalent to almost half a football field, being approximately 1.6 km deep [13]. Most of the Gulf floor is muddy, but the coral colonies that appear are vital oases for marine life in the cold ocean depths. There is still a great connection between animals that live on the surface and the life in the deep-sea, and problems in this ecosystem can cause damage to the entire marine environment for a long time.

The largest and richest coral reef on the planet, the Great Barrier Reef in Australia, has also been suffering from the leakage of oil and its derivatives. At the same time as the Gulf of Mexico disaster, in April 2010, the Chinese ship Shen Neng 1 ran aground near the barrier reef. The accident was responsible for a spill from 3 to 4 tons of oil, reaching about 3 km in length and 250 m in width, the largest ever recorded in the region. Some damaged areas have become completely devoid of marine life. It is estimated that this disaster will cause considerable damages to the coral reef in the long term, lasting for approximately 20 years [14].

In 2019, more than $5 \times 10^3$ tons of crude oil leaked from the Greek oil tanker Bouboulina reached more than 3000 km of the Brazilian coast and more than 980 beaches. The oil reached important coral reefs, including two of the largest protected coral reef areas in the South Atlantic: Costa dos Corais Environmental Protection Area and the Abrolhos Marine National Park [15]. The Abrolhos Marine National Park contains the largest and richest coral reefs in the South Atlantic, with a high degree of endemism [16]. However, they are vulnerable to local and global impacts [17].

This oil spill was the most extensive and severe environmental disaster ever recorded in Brazilian history [15]. However, environmental monitoring studies are needed to assess the real impact of this oil spill on marine life in the affected coral reefs.

## 3. Effects of oil on coral reefs

Coral reefs are among the most biodiverse and productive ecosystems on Earth, with great ecological and economic importance. However, they are especially sensitive to environmental changes and vulnerable to impacts from various anthropogenic activities, including contamination by oil spills [18].

Direct contact with oil or its soluble fraction is extremely harmful to reefs, leading to the death of corals and, consequently, several other marine organisms. Stress caused by oil contamination on corals includes histological effects, such as tissue death; biochemical effects, such as changes in the zooxanthellae primary productivity and increase in mucus production; reproductive effects, such as larval premature expulsion and losses in larval settlement; development effects, such as changes in the calcification rates and muscle atrophy; and changes in the coral's symbiotic microbiota (**Table 1**).

A common feature of corals against petroleum contamination is their ability for biological accumulation. The hydrocarbons rapidly bioaccumulate in the tissues (probably due to the high concentration of lipids), where they undergo a slow degradation. This bioaccumulation can still expose the zooxanthellae to hydrocarbons. In areas with chronic contamination, the hydrocarbons remain for a long period in the deposited coral's exoskeleton. It is important to note that corals have the ability to acquire hydrocarbons from the water column. It has already been observed that, in some cases, hydrocarbons may be absent in the sediment but present in the coral skeleton. On the other hand, bioaccumulation allows the use of coral skeletons as historical records of hydrocarbons contaminations.

| DAMAGE TO CORAL | References |
|---|---|
| Tissue death | [19–23] |
| Impaired feeding | [20, 22, 24] |
| Impaired polyp retraction | [21, 22, 25, 26] |
| Impaired debug or recovery ability | [27, 28] |
| Increased mucus production | [22, 29–32] |
| Change in calcification rate | [21, 33–35] |
| Damage to the gonads | [29, 36] |
| Premature expulsion of larva | [37, 38] |
| Larval death | [36, 39, 40] |
| Impaired larval settlement | [36, 41–45] |
| Expulsion of zooxanthellae | [21, 29] |
| Change in primary production of zooxanthellae | [21, 23, 26, 46] |
| Increased expression of genes associated with oxidative, immune, and metabolic stress responses | [28, 47] |
| Muscle atrophy | [29] |
| Tissue and fragment loss | [48] |
| Bioaccumulation | [10, 49] |
| Change in the symbiotic microbiota | [23, 46] |

**Table 1.**
*Effects of oil pollution on corals.*

The differences in tolerance by coral species are an important consideration, and species morphology seems to be directly linked to its susceptibility. For example, studies have found that branched corals appear to be among the most susceptible, while massive corals are more tolerant against oil exposure, probably due to higher oil adhesion and tissue damage in the former ones [50, 51]. In a field study of the incident that occurred in Bahia Las Minas (Panamá), it was observed that almost all branched corals were dead in the reef areas where the spill occurred [39].

In addition, studies have shown that chronic contaminations, even with low oil concentrations, can be more harmful to corals than shorter contaminations with higher oil concentrations. Exposure to chronic oil contamination can impair coral's biological functions, such as reproduction and recruitment, which will considerably reduce the survival of the corals.

## 4. Remediation technologies

Despite the great risk of oil exploitation to coral reefs, there is still no efficient, sustainable and large-scale applicable remediation strategy to protect or to clean up reefs impacted by oil spills.

Three main groups of emergency response strategies have been used to minimize marine oil contamination: physical/mechanical methods, chemical methods, and biological methods.

The most widely used chemical method is the application of dispersants. Dispersant consists of a mixture of surfactants and solvents that act to break oil into small droplets that are dispersed into the water column, where they are subjected to natural processes – such as wind, waves, and currents – leading to the intensification of evaporation, dispersion, aggregation into marine oil snow, sedimentation and, also, biodegradation. An extensive discussion about environmental implications of dispersants use as oil spills response method can be found in [2].

Meanwhile, studies have shown that dispersants can be more harmful to corals than oil itself, not only because of its toxicity but also because of the consequent increase in hydrocarbon concentration throughout the water column [46, 47]. In fact, regardless of the type of oil (light, medium, or heavy crude oil), the NOAA Office of Response and Restoration cites, at least in shallow coral reef areas, the natural recovery as the lower impact alternative for oil spill response, due to the negative effect of dispersants in those organisms [52]. In this sense, the use of dispersants in emergency response plans is usually avoided in sensitive areas and even subjected to environmental trade-off analysis, usually done by *Spill Impact Mitigation Assessment* (SIMA) methodology (see [53]). Nevertheless, the application of dispersants is mandatory in some extreme cases (i.e., blowout), for oil spill control or explosivity risk reduction, even in sensitive areas. Among examples of disasters where dispersants were prioritized, those occurred in Bahia Las Minas in Panama, near the Great Barrier Reef in Australia and the explosion of the Deepwater Horizon platform in the Gulf of Mexico must be highlighted.

A methodology that has shown excellent results in oil degradation and in reducing coral mortality is microbial bioremediation in conjunction with the use of probiotic microorganisms for corals.

The microorganisms associated with corals (coral microbiome) include virus, dinoflagellates, archaea, bacteria, and fungi, which are essential for the host's health [54–56]. These microorganisms provide multiple beneficial functions, such as production of photosynthate, supply of other micronutrients, protection against pathogens, nitrogen fixation, and UV-damage protection [57–64].

The coral animal and its associated microorganisms are known as coral holobiont. This holobiont is a dynamic system, whose members can fluctuate depending on environmental conditions and daily requirements [65]. The microbiome interactions can drive holobiont biology and define its phenotype [66].

Bioremediation is a biological remediation method that uses living organisms, or their enzymes, to reduce or remove environmental contaminants, and it can be applied in a wide range of ecosystems. Microorganisms are directly involved in biogeochemical cycles as key drivers of the degradation of many carbon sources, including petroleum hydrocarbons. Several microorganisms, such as bacteria, cyanobacteria, green algae, and fungi, can degrade different components of petroleum under different environmental conditions [67]. The use of probiotic microorganisms for corals has recently been proposed as a promising method to improve coral health, potentially promoting coral resistance and resilience helping to protect and recover impacted reefs [23, 54, 68–71]. The probiotic microorganisms can enhance coral fitness through their symbiotic relationships with the host [72–76].

In sensitive ecosystems such as coral reefs, bioremediation in conjunction with the probiotics for corals represents an efficient, sustainable, and low-cost alternative to chemical remediation of oil spills [23, 46].

Bioremediation can be carried out through bioaugmentation or biostimulation. The first one is the strategy in which pollutant degrading microorganisms, previously selected in laboratory are introduced into the contaminated environment. The use of microorganisms from the same environment in which they will be introduced is not mandatory; however, it is recommended. Biostimulation, on the other hand, is a bioremediation strategy that aims to stimulate the ability of native microorganisms to degrade pollutants, by identifying and adjusting certain physical and chemical parameters that may be undermining their biodegradation rate.

Studies based on bioaugmentation have shown that it is possible to manipulate the microbiome of coral species, making corals more resistant to environmental changes, and impacts. Some authors [23] showed that using probiotic and oil-degrading microorganisms to improve the health of corals under stress can protect corals from the effects of oil exposure. In this study, a consortium of 10 different bacterial species capable of degrading oil was isolated from corals of the *Mussismilia harttii* species. This bacterial consortium was able to degrade 72.75% of total petroleum hydrocarbons (TPH), and 56.20% of 38 polycyclic aromatic hydrocarbons (PAHs) in only 10 days of the experiment. Besides, the consortium was able to preserve the photochemical ability of zooxanthellae, protecting the corals against the negative effect of oil exposition.

Some authors [46] selected a multi-domain microbial consortium composed of bacteria, filamentous fungi, and yeast, which was able to mitigate oil impacts, substantially degrading oil components and improving coral health in the presence of oil. The microbial consortium was isolated from seawater and from corals of the species *Millepora alcicornis* and *Siderastrea stellata*. The coral species evaluated was *M. alcicornis*, a different species from that evaluated by other authors [23]. These studies showed that using probiotic and oil-degrading microorganisms to improve the health of corals under stress from oil exposure can foster coral survival. This bioremediation strategy can help companies and government agencies about the use of chemical or biological remediation since it has been shown that probiotic and oil-degrading microorganisms minimize the negative effects of oil without being toxic to coral.

In this context, some authors [54] proposed the term BMC (Beneficial Microorganisms for Corals) to refer to probiotic coral bacteria used to increase overall coral fitness through specific mechanisms, and suggested strategies for the use of this knowledge to manipulate the microbiome to restore and protect coral reefs. These strategies can be used not only to mitigate the stress and impacts of toxic compounds such as oil spills, but also to promote coral nutrition and growth, deter pathogens, and benefit early life-stage development.

## Author details

Luanny Fernandes[1], Flávia L. Carmo[2], Hugo E. de Jesus[1], Adriana U. Soriano[3] and Henrique F. Santos[1*]

1 Department of Marine Biology, Fluminense Federal University - UFF, Niterói, RJ, Brazil

2 Institute of Microbiology Paulo de Góes, Federal University of Rio de Janeiro – UFRJ, Rio de Janeiro, RJ, Brazil

3 Leopoldo A. Miguez de Mello Research and Development Center, CENPES, PETROBRAS -Petróleo Brasileiro S.A, Rio de Janeiro, RJ, Brazil

*Address all correspondence to: henriquefds@id.uff.br

## References

[1] Adipah S. Introduction of petroleum hydrocarbons contaminants and its human effects. Journal of Environmental Science and Public Health. 2018;**03**(01):1-9. DOI: 10.26502/jesph.96120043

[2] NASEM - National Academies of Sciences Engineering and Medicine. The use of dispersants in marine oil spill response. 2019. DOI: 10.17226/25161

[3] Fingas M, Brown CE. A review of oil spill remote sensing. Sensors (Switzerland). 2018;**18**:91. DOI: 10.3390/s18010091

[4] Southward AJ, Southward EC. Recolonization of rocky shores in Cornwall after use of toxic dispersants to clean up the Torrey Canyon Spill. Journal of the Fisheries Research Board of Canada. 1978;**35**(5):682-706. DOI: 10.1139/f78-120

[5] Peterson CH, Rice SD, Short JW, Esler D, Bodkin JL, Ballachey BE, et al. Long-term ecosystem response to the Exxon Valdez Oil Spill. Science. 2003;**302**(5653):2082-2086. DOI: 10.1126/science.1084282

[6] Atlas RM, Hazen TC. Oil biodegradation and bioremediation: A tale of the two worst spills in U.S. history. Environmental Science & Technology. 2011;**45**(16):6709-6715. DOI: 10.1021/es2013227

[7] Villela HDM, Peixoto RS, Soriano AU, Carmo FL. Microbial bioremediation of oil contaminated seawater: A survey of patent deposits and the characterization of the top genera applied. Science of the Total Environment. 2019;**666**:743-758. DOI: 10.1016/j.scitotenv.2019.02.153

[8] Farrington JW. Oil pollution in the marine environment I: Inputs, big spills, small spills, and dribbles. Environment: Science and Policy for Sustainable Development. 2013;**55**(6):3-13. DOI: 10.1080/00139157.2013.843980

[9] Keramea P, Spanoudaki K, Zodiatis G, Gikas G, Sylaios G. Oil spill modeling: A critical review on current trends, perspectives, and challenges. Journal of Marine Science and Engineering. 2021;**9**:1-41

[10] Burns KA, Knap AH. The Bahia Las Minas oil spill hydrocarbon uptake by reef building corals. Marine Pollution Bulletin. 1989;**20**(8):391-398. DOI: 10.1016/0025-326X(89)90317-2

[11] Vogt IP. Coral reefs in Saudi Arabia: 3.5 years after the Gulf War oil spill. Coral Reefs. 1995;**14**(4):271-273. DOI: 10.1007/BF00334351

[12] Bak RPM. Effects of chronic oil pollution on a Caribbean coral reef. Marine Pollution Bulletin. 1987;**18**(10):534-539. DOI: 10.1016/0025-326X(87)90537-6

[13] Fisher CR, Hsing P-Y, Kaiser CL, Yoerger DR, Roberts HH, Shedd WW, et al. Footprint of Deepwater Horizon blowout impact to deep-water coral communities. Proceedings of the National Academy of Sciences. 2014;**111**(32):11744-11749. DOI: 10.1073/pnas.1403492111

[14] Bateman S. Coastal state regulation of navigation in adjacent waters - the example of the torres strait and great barrier reef. Proceedings of the Advisory Board on the Law of the Sea (ABLOS) Conference; 2010

[15] de Oliveira SM, Teixeira CEP, Bezerra LEA, Paiva SV, Tavares TCL,

Garcia TM, et al. Oil spill in South Atlantic (Brazil): Environmental and governmental disaster. Marine Policy. 2020;**115**:103879. DOI: 10.1016/j.marpol.2020.103879

[16] Werner TB, Pinto LP, Dutra GF, do Prado Pereira PG. Conserving the southern Atlantic's richest coastal biodiversity into the next century. Coastal Management. 2000;**28**(1):99-108. DOI: 10.1080/089207500263684

[17] Duarte GAS, Villela HDM, Deocleciano M, Silva D, Barno A, Cardoso PM, et al. Heat waves are a major threat to turbid coral reefs in Brazil. Frontiers in Marine Science. 2020;**7**:179. DOI: 10.3389/fmars.2020.00179

[18] Turner NR, Renegar DA. Petroleum hydrocarbon toxicity to corals: A review. Marine Pollution Bulletin. 2017;**119**:1-16

[19] Johannes RE. Coral reefs and pollution. In: Ruivo M, editor. Marine Pollution and Sea Life. Fishing News (Books): London; 1972. pp. 364-374

[20] Reimer AA. Effects of crude oil on corals. Marine Pollution Bulletin. 1975;**6**(3):39-43. DOI: 10.1016/0025-326X(75)90297-0

[21] Neff JM, Anderson JW. Response of marine animals to petroleum and specific petroleum hydrocarbons. New York: Halstead Press; 1981:1-177

[22] Wyers SC, Frith HR, Dodge RE, Smith SR, Knap AH, Sleeter TD. Behavioural effects of chemically dispersed oil and subsequent recovery in Diploria strigosa (Dana). Marine Ecology. 1986;**7**(1):23-42. DOI: 10.1111/j.1439-0485.1986.tb00146.x

[23] Santos HF, Duarte GAS, Rachid CTDC, Chaloub RM, Calderon EN, Marangoni LFDB, et al. Impact of oil spills on coral reefs can be reduced by bioremediation using probiotic microbiota. Scientific Reports. 2015;**5**:1-11. DOI: 10.1038/srep18268

[24] Lewis JB. Effect of crude oil and an oil-spill dispersant on reef corals. Marine Pollution Bulletin. 1971;**2**(4):59-62. DOI: 10.1016/0025-326X(71)90211-6

[25] Elgershuizen JHBW, de Kruijf HAM. Toxicity of crude oils and a dispersant to the stony coral Madracis mirabilis. Marine Pollution Bulletin. 1976;**7**(2):22-25. DOI: 10.1016/0025-326X(76)90305-2

[26] Cook CB, Knap AH. Effects of crude oil and chemical dispersant on photosynthesis in the brain coral Diploria strigosa. Marine Biology. 1983;**78**(1):21-27. DOI: 10.1007/BF00392967

[27] Bak RPM, Elgershuizen JHBW. Patterns of oil-sediment rejection in corals. Marine Biology. 1976;**37**(2):105-113. DOI: 10.1007/BF00389121

[28] DeLeo DM, Glazier A, Herrera S, Barkman A, Cordes EE. Transcriptomic responses of deep-sea corals experimentally exposed to crude oil and dispersant. Frontiers in Marine Science. 2021;**8**:308. DOI: 10.3389/fmars.2021.649909

[29] Peters EC, Meyers PA, Yevich PP, Blake NJ. Bioaccumulation and histopathological effects of oil on a stony coral. Marine Pollution Bulletin. 1981;**12**(10):333-339. DOI: 10.1016/0025-326X(81)90106-5

[30] White HK, Hsing PY, Cho W, Shank TM, Cordes EE, Quattrini AM, et al. Impact of the deepwater horizon oil spill on a deep-water coral community in the Gulf of Mexico. Proceedings of the National Academy of Sciences of the United States of America. 2012;**109**:20303-20308

[31] Buskey EJ, White HK, Esbaugh AJ. Impact of oil spills on marine life in the Gulf of Mexico: Effects on plankton, nekton, and deep-sea benthos. Oceanography. 2016;**29**(3):174-181. DOI: 10.2307/24862719

[32] Santos HF, Santos LF, Jesus HE, Hf Lacerda C, Mies M. The South Atlantic coral Mussismilia harttii actively and quickly removes heavy crude oil from its surface. Bulletin of Marine Science. 2020;**96**(4):803-804. DOI: 10.5343/803

[33] Dodge RE, Wyers SC, Frith HR, Knap AH, Smith SR, Sleeter TD. The effects of oil and oil dispersants on the skeletal growth of the hermatypic coral Diploria strigosa. Coral Reefs. 1984;**3**(4):191-198. DOI: 10.1007/BF00288254

[34] Guzman HM, Jackson JBC, Weil E. Short-term ecological consequences of a major oil spill on Panamanian subtidal reef corals. Coral Reefs. 1991;**10**(1):1-12. DOI: 10.1007/BF00301900

[35] Glynn PW, Colley SB, Eakin CM, Smith DB, Cortés J, Gassman NJ, et al. Reef coral reproduction in the eastern Pacific: Costa Rica, Panamá, and Galápagos Islands (Ecuador) II Poritidae. Marine Biology. 1994;**118**(2):191-208. DOI: 10.1007/BF00349785

[36] Rinkevich B, Loya Y. Laboratory experiments on the effects of crude oil on the Red Sea coral Stylophora pistillata. Marine Pollution Bulletin. 1979;**10**(11):328-330. DOI: 10.1016/0025-326X(79)90402-8

[37] Cohen Y, Nissenbaum A, Eisler R. Effects of Iranian crude oil on the red sea octocoral heteroxenia fuscescens. Environmental Pollution 1977. 1970;**12**(3):173-186. DOI: 10.1016/0013-9327 (77)90051-9

[38] Loya Y, Rinkevich B. Abortion effect in corals induced by oil pollution. Marine Ecology Progress Series. 1979;**1**(1):77-80

[39] Yender RA, Michel J. Oil toxicity to corals. In: Oil Spills in Coral Reefs: Planning & Response Considerations. Seattle, Washington: US Department of Commerce, National Oceanic and Atmospheric Administration, National Ocean Service, Office of Response and Restoration; 2010;**2**:25-35

[40] Toyoshima J, Nakamura N, Fujii I, Watanabe A. Environmental impact of oil spill incidents on coral reefs and mangrove forests, and monitoring. OPRI Perspectives. 2020;**13**:1-18

[41] Te FT. Effects of two petroleum products on Pocillopora damicornis Planulae. Pacific Science. 1991;**45**(3):290-298

[42] Kushmaro A, Henning G, Hofmann DK, Benayahu Y. Metamorphosis of Heteroxenia fuscescens planulae (Cnidaria: Octocorallia) is inhibited by crude oil: A novel short term toxicity bioassay. Marine Environmental Research. 1997;**43**(4):295-302. DOI: 10.1016/S0141-1136(96)00092-X

[43] Goodbody-Gringley G, Wetzel DL, Gillon D, Pulster E, Miller A, Ritchie KB. Toxicity of deepwater horizon source oil and the chemical dispersant, Corexit® 9500, to coral larvae. PLoS One. 2013;**8**(1):e45574. DOI: 10.1371/journal.pone.0045574

[44] Epstein N, Bak RPM, Rinkevich B. Toxicity of third generation dispersants and dispersed Egyptian crude oil on Red Sea coral larvae. Marine Pollution Bulletin. 2000;**40**(6):497-503. DOI: 10.1016/S0025-326X(99)00232-5

[45] Negri AP, Heyward AJ. Inhibition of fertilization and larval metamorphosis

of the coral Acropora millepora (Ehrenberg, 1834) by petroleum products. Marine Pollution Bulletin. 2000;**41**(7-12):420-427. DOI: 10.1016/S0025-326X(00)00139-9

[46] Silva DP, Villela HDM, Santos HF, Duarte GAS, Ribeiro JR, Ghizelini AM, et al. Multi-domain probiotic consortium as an alternative to chemical remediation of oil spills at coral reefs and adjacent sites. Microbiome. 2021;**9**(1):1-19. DOI: 10.1186/s40168-021-01041-w

[47] DeLeo DM, Ruiz-Ramos D, v., Baums IB, Cordes EE. Response of deep-water corals to oil and chemical dispersant exposure. Deep-Sea Research Part II: Topical Studies in Oceanography. 2016;**129**:137-147. DOI: 10.1016/j.dsr2.2015.02.028

[48] Montagna PA, Girard F. Deep-Sea benthic faunal impacts and community evolution before, during, and after the Deepwater Horizon Event. In: Deep Oil Spills. Cham: Springer International Publishing; 2020. pp. 355-373

[49] Readman JW, Tolosa I, Law AT, Bartocci J, Azemard S, Hamilton T, et al. Discrete bands of petroleum hydrocarbons and molecular organic markers identified within massive coral skeletons. Marine Pollution Bulletin. 1996;**32**(5):437-443. DOI: 10.1016/0025-326X(96)83974-9

[50] Turner NR, Renegar DA. Petroleum hydrocarbon toxicity to corals: A review. Marine Pollution Bulletin. 2017;**119**:1-16

[51] Renegar ADA, Schuler P, Turner N, Dodge R, Riegl B, Knap A, et al. TROPICS field study (Panama), 32-year site visit: Observations and conclusions for near shore dispersant use NEBA and trade-offs. International Oil Spill Conference Proceedings. 2017;**2017**(1):3030-3050. DOI: 10.7901/2169-3358-2017.1.3030

[52] Hale C, Graham L, Maung-Douglass E, Sempier S, Skelton T, Swann L, Wilson M. Corals and Oil Spills. Oil Spill Science - Sea Grant Programs of the Gulf of Mexico. 2017. Available from: http://masgc.org/oilscience/oil-spill-science-corals.pdf

[53] IPIECA. Guidelines on implementing spill impact mitigation assessment (SIMA). IOGP Report 593. 2017. Available from: https://www.ipieca.org/resources/awareness-briefing/guidelines-on-implementing-spill-impact-mitigation-assessment-sima/

[54] Peixoto RS, Rosado PM, Leite DCA, Rosado AS, Bourne DG. Beneficial microorganisms for corals (BMC): Proposed mechanisms for coral health and resilience. Frontiers in Microbiology. 2017;**8**:341. DOI: 10.3389/fmicb.2017.00341

[55] Rohwer F, Seguritan V, Azam F, Knowlton N. Diversity and distribution of coral-associated bacteria. Marine Ecology Progress Series. 2002;**243**:1-10. DOI: 10.3354/meps243001

[56] Rosenberg E, Koren O, Reshef L, Efrony R, Zilber-Rosenberg I. The role of microorganisms in coral health, disease and evolution. Nature Reviews Microbiology. 2007;**5**:355-362

[57] Janouškovec J, Horák A, Barott KL, Rohwer FL, Keeling PJ. Global analysis of plastid diversity reveals apicomplexan-related lineages in coral reefs. Current Biology. 2012;**22**(13):R518-R519. DOI: 10.1016/j.cub.2012.04.047

[58] Kimes NE, van Nostrand JD, Weil E, Zhou J, Morris PJ. Microbial functional structure of *Montastraea faveolata*, an important Caribbean reef-building coral,

differs between healthy and yellow-band diseased colonies. Environmental Microbiology. 2010;**12**(2):541-556. DOI: 10.1111/j.1462-2920.2009.02113.x

[59] Raina JB, Tapiolas D, Willis BL, Bourne DG. Coral-associated bacteria and their role in the biogeochemical cycling of sulfur. Applied and Environmental Microbiology. 2009;**75**(11):3492-3501. DOI: 10.1128/AEM.02567-08

[60] Wegley L, Edwards R, Rodriguez-Brito B, Liu H, Rohwer F. Metagenomic analysis of the microbial community associated with the coral *Porites astreoides*. Environmental Microbiology. 2007;**9**(11):2707-2719. DOI: 10.1111/j.1462-2920.2007.01383.x

[61] Dunlap WC, Shick JM. Ultraviolet radiation-absorbing mycosporine-like amino acids in coral reef organisms: A biochemical and environmental perspective. Journal of Phycology. 1998;**34**(3):418-430

[62] Ritchie KB. Bacterial Symbionts of corals and Symbiodinium. In: Beneficial Microorganisms in Multicellular Life Forms. Berlin Heidelberg: Springer; 2012. pp. 139-150

[63] Fine M, Loya Y. Endolithic algae: An alternative source of photoassimilates during coral bleaching. Proceedings of the Royal Society B: Biological Sciences. 2002;**269**(1497):1205-1210. DOI: 10.1098/rspb.2002.1983

[64] Santos HF, Carmo FL, Duarte G, Dini-Andreote F, Castro CB, Rosado AS, et al. Climate change affects key nitrogen-fixing bacterial populations on coral reefs. ISME Journal. 2014;**8**(11):2272-2279. DOI: 10.1038/ismej.2014.70

[65] Thompson JR, Rivera HE, Closek CJ, Medina M. Microbes in the coral holobiont: Partners through evolution, development, and ecological interactions. Frontiers in Cellular and Infection Microbiology. 2014;**4**(DEC):176. DOI: 10.3389/fcimb.2014.00176

[66] Hernandez-Agreda A, Gates RD, Ainsworth TD. Defining the core microbiome in corals' microbial soup. Trends in Microbiology. 2017;**25**:125-140

[67] Peixoto RS, Vermelho AB, Rosado AS. Petroleum-degrading enzymes: Bioremediation and new prospects. Enzyme Research. 2011:1-7. DOI: 10.4061/2011/475193

[68] Damjanovic K, Blackall LL, Webster NS, van Oppen MJH. The contribution of microbial biotechnology to mitigating coral reef degradation. Microbial Biotechnology. 2017;**10**(5):1236-1243. DOI: 10.1111/1751-7915.12769

[69] Sweet M, Ramsey A, Bulling M. Designer reefs and coral probiotics: Great concepts but are they good practice? Biodiversity. 2017;**18**:19-22

[70] Rosado PM, Leite DCA, Duarte GAS, Chaloub RM, Jospin G, Nunes da Rocha U, et al. Marine probiotics: Increasing coral resistance to bleaching through microbiome manipulation. ISME Journal. 2019;**13**(4):921-936. DOI: 10.1038/s41396-018-0323-6

[71] Santoro EP, Borges RM, Espinoza JL, Freire M, Messias CSMA, Villela HDM, et al. Coral microbiome manipulation elicits metabolic and genetic restructuring to mitigate heat stress and evade mortality. Science Advances. 2021;**7**(33):eabg3088. Available from: https://www.science.org

[72] Burriesci MS, Raab TK, Pringle JR. Evidence that glucose is

the major transferred metabolite in dinoflagellate-cnidarian symbiosis. Journal of Experimental Biology. 2012;**215**(19):3467-3477. DOI: 10.1242/jeb.070946

[73] Davy SK, Allemand D, Weis VM. Cell biology of cnidarian-dinoflagellate symbiosis. Microbiology and Molecular Biology Reviews. 2012;**76**(2):229-261. DOI: 10.1128/mmbr.05014-11

[74] Tremblay P, Grover R, Maguer JF, Legendre L, Ferrier-Pagès C. Autotrophic carbon budget in coral tissue: A new 13C-based model of photosynthate translocation. Journal of Experimental Biology. 2012;**215**(8):1384-1393. DOI: 10.1242/jeb.065201

[75] Alagely A, Krediet CJ, Ritchie KB, Teplitski M. Signaling-mediated cross-talk modulates swarming and biofilm formation in a coral pathogen *Serratia marcescens*. ISME Journal. 2011;**5**(10):1609-1620. DOI: 10.1038/ismej.2011.45

[76] Kvennefors ECE, Sampayo E, Kerr C, Vieira G, Roff G, Barnes AC. Regulation of bacterial communities through antimicrobial activity by the coral holobiont. Microbial Ecology. 2012;**63**(3):605-618. DOI: 10.1007/s00248-011-9946-0

Chapter 4

# Perspective Chapter: Electric Reefs Enhance Coral Climate Change Adaptation

*Thomas J. F. Goreau*


## Abstract

Survival of coral reefs from globally rising temperature, sea level, and pollution requires new regeneration methods that greatly increase coral growth rates, survival, and resistance to accelerating extreme environmental stresses. Electric ("Biorock") limestone reefs can be grown in any size or shape, get stronger with age, grow back if damaged, and increase settlement, growth, survival, and stress resistance of all marine organisms examined. Electrified corals survive repeated severe bleaching events when over 90% of corals on nearby reefs die. Electric reefs are open mesh frameworks with more vertical levels of holes and surfaces than natural reefs, which can be grown in places where natural reefs cannot grow due to lack of substrate or unsuitable physical and chemical conditions. They strongly enhance reef physical structure, wave absorption, ecological function, biodiversity, productivity, and habitat and ecosystem services including shore protection, sand generation, and fisheries habitat, even at severely degraded sites where no natural regeneration takes place. Electric reefs, optimized for local needs, provide superior cost-effective options for shore protection, conservation, sustainable mariculture, and regeneration of coastal ecosystems that grow to keep pace with sea level rise. They will provide an essential tool to develop sustainable "Blue Economies."


**Keywords:** coral reefs, structure and function, space frame structures, wave absorption, water throughflow, ecosystem services, climate change, adaptation, electrolysis, electric reefs, biorock, blue carbon

## 1. Introduction

Coral reef structure and function are imminently threatened with extinction from pollution, global warming, and pathogenic diseases, problems that began to affect coral reefs in the 1980s [1, 2].

Key parameters of coral reef structure and function based on interactions with water flow include the following: 1) how efficiently reefs absorb wave energy, 2) how reef water throughflow prevents scouring erosion in front from wave pressure buildup and reflection, and 3) how quickly water flows through the reef structure, providing food and removing wastes [1]. All are highly sensitive to three-dimensional solid reef structure [3] and also to what is defined here as "reef negative structure," the size, shape, and interconnections of empty water-filled spaces in the reef framework, commonly called holes, cracks, crevices, and caves that result in reef porosity and permeability to throughflow. Here, we discuss how structural benefits of reef throughflow can be greatly increased using Biorock electric reef technology to adapt to accelerating global warming and global sea level rise threats to coastal ecosystems and infrastructure.

Corals have evolved superb adaptation to wave forces and flow, so reef morphology at every scale elegantly recapitulates the wave forces on them. Charles Darwin realized this when he climbed Mount Orohena, Tahiti's highest ancient volcanic peak, to admire the smooth curves of coral reefs completely encircling the islands of Tahiti and Moorea, except for gaps wherever rivers poured freshwater and mud into the sea. Darwin recognized the key features of coral reefs: They were the most effective wave energy absorbers, built beaches, islands, and atolls, had the greatest marine biodiversity, needed warm, shallow, clear waters, and were damaged by freshwater and mud from land. There is no evidence Darwin ever went into the water himself, his insights came from studying Admiralty Nautical Charts (British Royal Navy maps of where ships could run aground, based on meticulous bottom soundings), from leaning over the boat railing when the water was calm and clear at anchor, and rare glimpses through an "underwater telescope," the bottom viewing glass sailors used to see if anchors set properly on the bottom. His focus was on coral as geological island builders, and he never looked at coral reefs closely enough to gain evolutionary insights from them. Darwin knew corals could be propagated by fragmentation on clean bottoms from information provided by a Jamaican geologist about reefs in Panama, and from a Scotsman in Madagascar, not from his own personal observations [4]. Although coral fragmentation and propagation have been routinely used by coral scientists ever since, it has recently been claimed as a "new discovery."

The other co-discoverer of evolution, Alfred Wallace, saw coral reefs of Ambon Bay in Indonesia by looking over the side of a boat at port and was mesmerized by the vast untouchable universe beneath the waves, which he could only glimpse ephemerally but never touch, with shapes and functions even more complex, varied, and intricate than the astonishing tropical jungle life he studied [5]. The blind seer of Ambon, Georg Eberhard Rumpf (Rumphius) described the skeletons of corals and shells by touch in the 1600s [6]. Now, those corals have vanished due to mud and pollution, and Biorock Indonesia electric reefs are growing the last corals in Ambon Bay (https://www.globalcoral.org/biorock-brings-corals-back-in-ambon/). After development of diving as a research tool at the 1947 Bikini Atoll Scientific Resurvey, scientists began to dive into reefs to learn how corals functioned as habitat builders from shallow to deep [1].

## 2. Collapse of coral reef structure and function

The priceless ecosystem services of coral reefs may soon be lost unless methods are found to regenerate reefs in the ocean that are more resistant to high temperature and pollution stresses. Coral reef ecosystem services per unit area are the most valuable of any ecosystem on earth: so valuable that economic losses caused by coral reef destruction alone make up nearly 60% of total global value of all of Earth's ecosystem service losses between 1997 and 2011 [7]. Those estimates of reef damage are obsolete because they were made before accelerated collapse of coral reefs from bleaching, diseases, and pollution, which happened in the record hotter years since that paper was written. Coral reefs occupy less than 0.1% of the ocean's surface, so all other ecosystems in the world, which cover more than 99.9% of the planet, contribute only 40% of all global economic losses from ecosystem deterioration. The economic losses per unit area in coral reefs are thousands of times greater than the global average, making clear the exceptional vulnerability of coral reefs to destruction, and resulting global economic impacts on shoreline protection, fisheries, tourism, and biodiversity.

Healthy coral reefs absorb up to 97% of incoming wave energy [8], and healthy corals grow back in years after being damaged, but only if water temperatures, sediments, nutrients, pollutants, pathogens, and wave energy are within the coral's tolerance range [1]. Dead reef structures are crumbling and collapsing from bioerosion by boring sponges, clams, worms, algae, fungi, and bacteria, which perforate dead coral skeletons and limestone with holes until seemingly solid reef rock eventually shatters and collapses in severe storm waves [1]. Once gone, shore protection dead coral reefs used to provide has to be replaced by concrete and stone walls costing tens of thousands of dollars per meter, which provide no ecosystem services such as fish habitat and sand production. Instead of generating sand like coral reefs do, concrete and rock walls cause scouring and sand erosion in front, and then underneath them, until they are undermined and collapse.

Spectacular elkhorn and staghorn coral reefs grew right up to the surface around Caribbean islands, so dense one could not swim over them [1]. These were dynamited in Bonaire and other places so divers could swim from the beach to reach the outer slope. Now, those reefs are almost entirely gone except for a few fragments of dead coral on the slopes below, and the beaches are washing away. There is urgent need worldwide to regenerate reef structures to prevent beach loss: Half the world's beaches are expected to vanish in this century [9], even if sea level rise, wave strength, and coral bleaching caused by global warming do not increase, as now seems certain to happen due to political failure to halt and reverse greenhouse gas emissions.

Unless coastal ecosystem collapse is reversed quickly, tropical coastal countries have little hope of developing sustainable Blue Economies. This will require active restoration with artificial structures where the reefs themselves have collapsed. All around the world dead reefs have disappeared, flattened, and pulverized to rubble by hurricanes, nearly barren of living corals and fish. Artificial reefs can regenerate some lost biomass, biodiversity, and ecosystem function in such places, but vary greatly in their effects depending on structural design and ambient conditions.

## 3. Types of artificial reefs

Artificial reefs can be classified by three generations of technological improvements. First-generation or "Indigenous" artificial reefs [10] are traditional fish habitats on the bottom or floating, constructed since ancient times by Indigenous Pacific Sea Peoples from natural rocks, bamboo, and coconut palm leaves to create Fish Aggregation Devices (FADs). Pacific fishermen know fish swarm beneath the shadow of floating tree trunks, so whenever they see a floating log, they head straight for it and fish under it as long as they can and still get back home. FADs anchored to rocks or deep moorings work so well that there is no point fishing elsewhere.

Indigenous bottom reefs can be astonishingly productive of fish and edible snails by creating habitat for desired species where there is little available. One such reef dismantled by fishermen in the Philippines yielded nearly a hundred rabbit fish and groupers in just a few square meters. These artificial reefs build up populations of some species, but do not restore habitat in the long run or on large scale because floating FADs made of coconut and bamboo rot and fall apart, and rock piles are deliberately dismantled to harvest fish and shellfish hiding inside, although usually reassembled for the next harvest a year or two later. In many places in Southeast Asia and the Pacific islands, you see ancient rock pile reefs along shores that had been regularly harvested for thousands of years, but which were abandoned after colonial administrators destroyed traditional resource use laws as a political threat to the new rulers.

Second-generation, or "modern" artificial reefs, are mostly made from massive exotic manufactured materials such as Portland Cement concrete, steel, ceramics, rubber, and plastics, and now, 3-D printing [10]. Larvae of coral reef organisms are particular about where they grow; they seek clean limestone surfaces on which to settle. Most coral reef organisms refuse to settle on exotic materials, so instead of hard corals found on limestone rock, artificial reef materials generate biological fouling communities dominated by slimy cyanobacteria, weedy algae, stinging hydroids, and bacteria eating sponges. These soft fouling communities are very different in structure, function, and biodiversity from true constructional coral reef communities. Fish will hide behind any structure where there are none, but artificial reefs made of exotic materials almost always fail to generate real coral reef ecosystems, and fish seem to prefer to aggregate in dead coral reefs than concrete. The individual modules from which these artificial reefs are constructed rust, corrode, collapse, crumble, and eventually fall apart, often leaving toxic materials behind. Or they are moved, removed, or destroyed by storm waves, making physical benefits temporary, and littering the sea floor with trash. Since hard coral settlement on their exotic surfaces is very sparse, second-generation artificial substrate reefs are based on propagation by coral fragmentation, which was well known to Charles Darwin, so we classify them as Pre-Darwinian reef restoration methods.

Pre-Darwinian artificial reefs have a very poor record of survival in storms and are often completely demolished, frequently without trace. After Hurricane Andrew hit South Florida, divers found all artificial reefs (mostly sunken ships) had moved from where they had been installed. The structures were ripped apart, with none, one, or many fragments being found far down-current. Despite lack of long-term structural stability, exotic material artificial reefs planted with fragmented coral clones are the fad of the moment with funding agencies, and hundreds of well-intentioned groups around the world are breaking corals and gluing them to concrete, rubber, plastic pipe, plastic fishing lines, and other underwater structures. These groups mean well, but have shown no long-term results because fragmented corals in these nurseries can grow in tanks with expensive pumps, lights, filters, and food with sufficient care and expense, but will die when put into the field as soon as the water gets too hot [11], muddy, polluted, or rough, and when disease strikes, these monoclonal coral plantations usually all die. Such methods are expensive and temporary solutions that typically last only until the first extreme stress event.

The well-known failures of these fragmentation reef regeneration methods in the face of global warming are reviewed elsewhere [11], so not repeated here. This chapter discusses optimization of Post-Darwinian reef structures to uniquely meet physical and biological challenges caused by anthropogenic stress to coral reef ecosystems from global climate change and pollution, which are increasing temperature, sea level, wave energy, nutrients, toxic pollution, and new epidemic diseases beyond the range corals glued to exotic material can tolerate.

Third-generation, or Post-Darwinian artificial reefs [10], grown in the sea by electrolysis [12], are a quantum leap ahead of all other artificial reefs, because they uniquely grow the natural limestone that coral reef organisms prefer to settle on [13–16]. Seawater electrolysis creates growing limestone reefs of any size or shape, which get stronger with age, and are self-repairing [15]. Electric reefs have survived category five hurricane impacts and severe coral bleaching events without damage [17]. Biorock reef structures show greatly increased settlement, growth, healing from damage, survival, resistance to severe environmental stress, and biodiversity of all marine organisms, preventing coral mortality from bleaching, absorbing wave energy, regenerating severely eroded beaches, and creating new habitats that allow selected valuable species to be grown at high density without added food, while absorbing $CO_2$ [14–16, 18].

The electrolysis process uses Safe Extra Low Voltage (SELV) direct current seawater electrolysis to grow limestone structures in the ocean, in any size or shape [12–16]. More than 600 electric reefs have been built since 1976 to ameliorate local marine ecosystem regeneration problems in some 45 countries around the world. All projects worked well if designed by a trained team, used the right materials and conditions, and were maintained. Untrained users of Biorock technology generally fail to get results shown in this chapter because of fundamental errors, reviewed in Ref. [15].

The principles of electric ecosystem regeneration have been learned by observation and experimentation, not derived from *a priori* physical principles or models. Indeed, the results were so unexpected that they could only have been learned from the observational natural history-based approach used by Rumphius, Linnaeus, Charles Darwin, Alfred Wallace, and Alexander Humboldt, and could not have been learned from conventional hypothesis-based research funding strategies.

## 4. Physical observations of electric reefs

There are no limits to the size or shape of reefs that can be grown in the ocean by electrolysis, and costs are far less than any Pre-Darwinian artificial reefs, while ecosystem service benefits are far greater [18]. The minerals produced by seawater electrolysis are the only marine construction materials that grow solid limestone structures, attach themselves solidly to the sea floor, "glue" or "cement" themselves to the bottom, and get stronger with age by growing new layers of limestone minerals from sea water. More remarkably, this material repairs itself, and the physically damaged areas grow back as long as they are electrically charged. Electrolysis produces solid materials at least 2–3 times harder than Portland Cement concrete that grow upward at rates of up to 2 centimeters per year, faster than global sea level rise, now about 0.3 centimeters per year [14]. Since they only grow while submerged, breakwaters made from this process match sea level rise and will not need to be constantly repaired, rebuilt, and re-topped to keep pace with rising waters like conventional marine construction. The oldest known rock breakwaters, built 7000 years ago to protect a village on the shores of the Eastern Mediterranean, failed when the ocean rose above it [19].

Electric reefs can be built in any form, as strong, porous, and permeable to water and sediment flow as desired. They are versatile and adaptive, and can evolve as needed by adding new modules to provide more wave absorption, or removing portions if more water flow is desired. For example, groynes can be made partially permeable to sand so they do not block longshore sand drift, causing erosion down-current like solid groynes. Sand permeability of such groynes can be adjusted as desired by adding or removing modules or segments when needed (https://blueregeneration.com/#).

Electric reefs are designed with interconnected open spaces and passages with high internal surface area that efficiently absorbs wave energy, while recycling nutrients on biological surfaces of organisms living on internal passage surfaces. Large open-framework structures have much higher negative space than massive reefs, allowing many times more water throughflow for their mass than real coral reefs. Open spaces are built into the structure at the start, while natural reefs are accretionary, adding on new corals only from the bottom up, until they fill the spaces between corals, so many holes are accidents of failure to grow or physical damage.

Because coral reefs grow from the bottom-up, they interact with water only by bottom friction. Once passages through the reef are filled in by solid massive corals and sediments, the reef is no longer porous and permeable to water flow, which is needed for efficient energy dissipation and internal carbon and nutrient recycling [1, 3]. As reef passages and pores fill in, reef structure negative space, porosity, and permeability decline, and so do biological growth, ecological function, and wave energy absorption capacity, as the reef reaches its upper vertical limits and transforms from an open growing structure to a solid eroding one. Blockage of holes through the reef prevents water throughflow and prevents wave pressures equilibrating across the reef structure, greatly increasing erosional forces on the reef front. Hole blockage causes wave reflection and sediment scour on the front surface, erosion that is minimized in reef frameworks with open passages. Conventional artificial reefs are usually made of solid rock and concrete, and so suffer severe scour erosion in front and underneath them in storm waves, while electric reefs nearby accumulate sand underneath them [17]. Some reinforced concrete structures have been designed with holes, but generally have poor water throughflow, and deteriorate as steel reinforcement rusts, expands, and cracks the concrete, the opposite of electric reefs that do not rust and gain strength with age.

Electric reefs are open cross-linked mesh frameworks that interact with water throughout their entire vertical height and cross section, not just the uppermost or outermost surfaces. Crucial design features of electric reefs are low surface area cross section perpendicular to flow, but high surface area parallel to flow, avoiding excessive drag stress forces that can tear solid structures apart, while providing internal surface area with tangential drag to efficiently slow down waves. They can be made with any number of layers of holes and passages, of any size or orientation, and so provide much more surface area and effective biological volume, with vastly less solid mass than a coral reef that accumulates solid material only from beneath. Different species of fishes, lobsters, octopus, oysters, giant clams, and other marine life are observed to prefer different size and shape holes for shelter and finding food, so structures can be designed to suit desired species. Biorock reef structures survived extreme hurricane waves that tore solid concrete structures apart (**Figure 1**, from [17]).

Mesh frame structures cover much more bottom area and volume with far less material than solid breakwaters, so they cost much less to build and install (**Figure 2**). A simple geodesic dome is shown in **Figure 2**. Despite low weight, these structures are extremely strong because of tensile interconnections and can be made in any shape to fit any bottom topography. Mesh frame structures are readily analyzed by finite element stress and strain analysis, which shows mesh structures made from varying length elements have superior mechanical properties to those made from uniformly sized elements [20]. Because such structures are largely open in their interiors, they allow flow of nutrients and food to marine organisms throughout the structure, and flushing of their wastes, resulting in elevated growth of all reef species throughout

**Figure 1.**
*Before and after the two worst hurricanes in Turks and Caicos Islands history. The biorock reef built up sand underneath, while most concrete artificial reefs were thrown out and concrete reef balls nearby suffered severe scour erosion and were buried beneath the sand (in background). Note swimmer for scale. Photographs by Lucy Wells.*

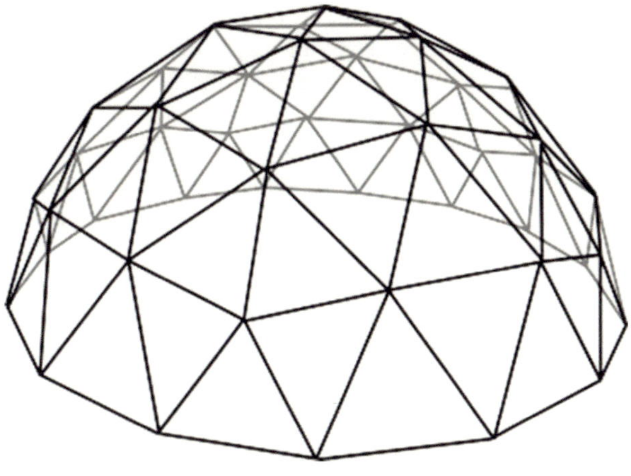

**Figure 2.**
*A geodesic dome space frame structure shows the basic principle of biorock reefs. This is a simple geodesic dome, but they can be made in any size or shape.*

the reefs negative space when compared with marine organisms living in cracks and crevices in solid frameworks, which are limited by slow transport of food and wastes. They provide much more biological surface area than breakwaters for wave energy absorption by friction and for food capture from the water by filter-feeding organisms such as corals and sponges.

**Figures 3–5** show the ranges of shapes of a small portion of around 150 electric reefs built by Biorock Indonesia at Pemuteran, Bali. Electric reef structures built on sand created lush reefs in a few years on bare sand sites that had been barren of corals and fishes. Electric reef structures on the dead reef, which was almost entirely killed by bleaching in 1998, stimulated tremendous spontaneous settlement of corals that turned nearby areas from only a few percent coral cover to nearly 100% live coral in around 5 years [15]. It is especially noteworthy that diversity of corals and fishes were greater 10 years after electrical regeneration than before bleaching. Soft corals (stress indicators) had covered much of the bottom before bleaching, but were replaced by hard corals after electric regeneration. More complex internal spaces in electric reefs created hiding places for large schools of fish, whose populations and diversity soared [16].

Electric reef hurricane performance was compared with conventional concrete artificial reefs in Grand Turk [17]. The electric reef was sitting loose on the bottom, not anchored, and had not been welded, only hand-tied together with wire. Thousands of corals were rescued from a reef where they were being killed by propeller wash sand and turbidity from tourist cruise ships. The corals had been attached loosely to the structure only a week before the two worst hurricanes in the island's history hit, just 3 days apart. Almost all houses on land were destroyed or damaged, but there was essentially no damage to corals or structures, although the electrical cable powering it was ripped out by waves. During the hurricanes sand built up under the electric reef and buried the lowest portions [17]. In sharp contrast, most concrete block artificial reefs inside the electric reef were washed out, and concrete reef balls on the sand nearby caused so much scour and erosion around and underneath them that they were undermined, dug their own graves, and sank partly into the sand or vanished completely beneath it (**Figure 1**). Electric reefs have shown their effectiveness at wave energy absorption by growing back severely eroded beaches at record rates [14, 21].

Electric reefs must be heavy and strong enough, or sufficiently well attached to the bottom, which they do not move or get damaged in extreme storms before growing strong. They can be attached to the ocean floor by drilling or with sand anchors. Electric reefs that survived Category 5 hurricanes were neither welded nor attached to the bottom in any way, and they sat on the bottom under their own weight and cemented themselves to it with growing limestone [17]. There was no damage to corals or structures on an electric reef on Saint Barthelemy on top of the dead reef flat in 1 meter depth of water. This reef flat absorbed the major force of huge breaking waves from Category 5 Hurricane Irma, which damaged and destroyed all buildings in the bay behind the reef (see video at https://www.globalcoral.org/biorock-electric-coral-reefs-survive-severe-hurricanes-little-no-damage/). The first electric reef built in Jamaica in the 1980s has been off power for more than 30 years, but is so solidly attached to the sea floor that it has survived every hurricane in three decades, and many of the original corals are still growing on it, although it no longer has enhanced growth or self-repair [16].

Electric reefs are weakest when first built and get steadily stronger, the opposite to all other construction materials, which are strongest when installed and deteriorate from that moment until they fail and collapse. The electrolysis process cements

**Figure 3.**
*A sample of the range of biorock coral habitats at the Karang Lestari project, Pemuteran, Bali, photographed in may 2012 by Euen Jae Im. These photos were all taken on 1 day, and they show only a small part of around 150 different biorock reefs at the site. Many reefs represent Balinese myths, designed by creative local artists. a) Four-year-old biorock reef on bare sand surrounded by a cloud of fish, zoned according to their preferred habitat. b) Four-year-old biorock reef on bare sand attracts schools of fishes. c) Biorock reefs attract spontaneous settlement by hard corals, soft corals, tunicates, sponges, crinoids, and juvenile fish, quickly building up highly biodiverse ecosystems. d) this biorock reef covers a lot of area with little material. By building up the height of the dead reef and attracting coral settlement, it has grown back a badly eroded beach on the shore behind it. e) Biorock mesh reefs stabilize loose rubble and can become quickly concealed under coral growth, armoring coastlines against erosion. f) a wide variety of fishes are attracted. Batfish are usually first to school around newly installed biorock reefs.*

sediments below the water surface, so is ideal for sub-bottom structural foundations, such as in Venice [22]. The bases of collapsing cliffs can be armored against bioerosion and loose rubble cemented together. In Indonesia, the Philippines, and Tanzania, baby corals settle on dead coral rubble of dynamited reefs, but die when waves flip them over. Electrolytic mineral growth on meshes stabilizes shifting substrate, and

(a) (b) (c) (d) (e) (f)

**Figure 4.**
*A sample of the range of biorock coral habitats at the Karang Lestari project, Pemuteran, Bali, photographed in may 2012 by Euen Jae Im. These photos were all taken on one day, and they show only a small part of around 150 different biorock reefs at the site. Many reefs represent Balinese myths, designed by creative local artists. a) this biorock reef contains three concentric domes, the inner one creating a darker cave-like habitat. b) Biorock reefs quickly build up dense local fish populations in barren areas. c) Biorock reefs attract cleaning shrimp and cleaner fish and are excellent places to watch fish behavior close up. d) these reefs are located on barren sand in front of a reef that was killed by bleaching in 1998. e) They brought back spectacular life to an area that had looked like a barren moonscape. Larger fish, such as groupers, prefer the dark interiors. f) the entire economy of the town or Pemuteran is built around ecotourism. People come from around the world to see the projects.*

coral fragments grow right up through it, allowing reefs to grow back in severely damaged sites where there had been no recovery.

Electrolysis can create reefs where there is no suitable bottom for reef growth, even in the deep sea. Floating electric reefs in a suburban Grand Bahama canal grew

**Figure 5.**
*A sample of the range of biorock coral habitats at the Karang Lestari project, Pemuteran, Bali, photographed in may 2012 by Euen Jae Im. These photos were all taken on 1 day, and they show only a small part of around 150 different biorock reefs at the site. Many reefs represent Balinese myths, designed by creative local artists. a) Biorock dome reef on sand. b) the first biorock reef, built in 2000, has spectacular growth of the blue coral, Heliopora coerula. Front side. c) Back of same reef, shaped like a Balinese fish trap called a Bumbung. It increased growth rate of corals on it 4.01 times more rapid than nearby controls, and stimulated massive spontaneous settlement of new corals all over the dead reef nearby that was killed by bleaching in 1998. d) Corals grow through and over biorock reefs, while fish swarm inside. e) Ingenious Balinese artists make spectacular reefs shaped like marine life, in this case a crab, in a new biorock reef just starting. f) Hybrid sculptures retell Balinese mythology.*

filter feeding reef communities dominated by sponges, hydroids, oysters, mussels, tunicates, and corals that helped filter and clean turbid waters, a method that could be used to help clean-polluted harbor waters. Floating solar-powered electric reefs can serve as superior Fish Aggregating Devices and offshore wave absorbers, increasing pelagic fish populations by providing shelter, shade, and food. Electric reefs can be used for floating offshore port installations [23]. The United Nations Sustainable Floating Cities Initiative, launched in April 2022, plans to use electric reefs on the bottom as mooring anchors and bottom fisheries habitat regeneration, and floating electric reefs for wave protection, mariculture, and cleaning polluted harbor waters.

Safe Extra Low Voltage (SELV) direct electrical currents are the cause of all physical, chemical, and biological benefits of the electrolysis process [1]. Limestone does not naturally form in sea water unless marine organisms spend their own biochemical energy growing shells and skeletons. The need for a trickle charge of electricity is also the method's major limitation, requiring reliable, cost-effective sources of electricity at sites where projects are needed. Extremely low voltages like those used for seawater electrolysis cannot be transmitted long distances except through large-expensive electrical cables due to power losses from voltage drop, so they are best provided by local electrical sources. Electric reefs can be powered by any source of electricity transformed into low-voltage direct current; we have used conventional alternating current, photovoltaic panels, windmills, ocean current and tidal energy turbines, wave energy generators, and batteries. Fuel cells can be used too and will likely become the method of choice for remote sites such as on the bottom of the deep ocean floor. Which power source to use depends on the choices available at each site, and their costs, so is a site-specific local design issue.

Electrolysis allows whole islands to be grown on shallow banks that are now being submerged by sea level rise, potentially allowing low island nations that are now abandoning inhabited islands due to erosion to adapt to climate change. Hard electrically grown limestone materials are permanent (the same material as the pyramids), eliminating need for frequent and costly repairs, or concerns that sea defenses will be eventually overtopped by global sea level rise. No other coastal defense technology evolves to match global sea level rise.

## 5. Biological observations of electric reefs

The physical benefits of structural wave attenuation and shelter exist independently of any biological benefits, but are greatly enhanced by them [16]. Most wave absorption capacity of electric reefs is provided by prolific biological growth over the structure, whose surface area greatly exceeds that of limestone chemically produced by electrolysis (**Figures 3–5**). Electric reefs grow and protect coastlines fastest where there are the best water quality conditions for healthy marine life; however, they also work even in sterile and toxic environments where little or no marine life can grow, though not as well. Wave-absorbing reefs to protect shorelines from erosion can be grown in waters whose physical, chemical, and biological conditions are too extreme or lethal for reef building organisms, for example, in severely polluted and anoxic harbors, or in front of Fukushima (where it would help trap radioactive Strontium-90 emissions). Electrolysis allows offshore reefs to be built along shallow coasts whether a reef was there before or not. Such reefs could save beaches lacking protective reefs that are replenished regularly at great expense by sand dumping "renourishment" that vanishes after the first big storm, such as Cancun or Miami Beach.

**Figures 3**-5 Examples of the huge variety of open framework electrolysis reefs. These photographs show only a small portion of around 150 electric reefs at the site, taken during a single day.

The electrolysis process provides direct biochemical energy benefits for health and resilience of marine organisms, empowering them to survive otherwise lethal conditions [14, 16]. Corals survive repeated severe bleaching on electric reefs, while corals away from them die, electrified intertidal saltmarsh grass grows deeper offshore than it could otherwise tolerate because of their prolific root growth, and sea grass has been grown on bare rock with no sediments, previously thought impossible [16]. Most seagrass, saltmarsh, and mangrove transplantation projects fail because plants are washed away by waves before roots can grow. Electrolysis regenerates natural coastal defenses and capability to adapt to sea level rise by accelerating plant growth both below and above ground, allowing marine plants to quickly establish themselves and proliferate by underground root growth. They can expand eroding coastal saltmarsh ecosystems seaward in places where they are now rapidly retreating inland because of increasing sea levels and storm strength, while increasing coastal "Blue Carbon" storage in living shorelines and sediment organic carbon in seagrasses, salt marshes, and mangroves [16].

Carrying capacities of reef species are determined by the size and shapes of holes available for their shelter and also for organisms they eat. Each species of octopus, lobster, crab, shrimp, and fish searches for holes of a certain shape to hide or searches for food. The number of suitable holes limits how many individuals fit any given area. Holes in a natural reef are not the result of systematic and abundant design, there may be only one suitable hole in a large area, and there is usually only one layer of holes on the bottom. Holes are often historical accidents limited to spaces not yet filled in by corals or sand, or cracks from structural failure where corals broke off and were washed out by extreme waves. Such holes are accidental, ephemeral, and temporary, filling in with coral, sand, or rubble as the reef grows. Electric frame mesh reefs have not just a few holes at the bottom like coral reefs do, but also have passages repeated up throughout the entire height of the structure, allowing larger populations of marine organisms to find shelter than possible on natural reefs, with better flow of water transporting food in and wastes out. Electric reefs of different shapes side by side build up populations of different fish species. Electric reefs can be designed for specific needs of desired species, once we learn what shape they prefer. Scores to hundreds of lobsters crowded into a few square meters of electric lobster habitat in Jamaica, Mexico, and Panama. Electric reefs have been completely overgrown with oysters, mussels, or corals that spontaneously settled [16]. High biodiversity of electric reefs allows the reef to internally generate its food supply for sustainable whole-ecosystem mariculture, without needing to add expensive manufactured food, avoiding most of the problems of monoclonal commercial mariculture [24]. A small sample of the range of electric reefs grown at one location, Pemuteran in Bali, is shown in **Figure 3**. They are more spectacular in video because of constant movement of fish schools around them, seen at: https://www.youtube.com/user/TheBiorockChannel.

Electrolysis not only amplifies reef physical and biological function, extending and regenerating them where it had been lost or degraded, it also allows shapes natural reefs do not have, and in locations where they cannot naturally exist. Floating electric reefs moored to the bottom can create upside down reefs in deep water, generating habitat for oceanic fish such as mahi-mahi and tuna. Electric reefs are predicted to be more effective than conventional Fish Aggregation Devices because they generate intense biological productivity and biodiversity, creating new food chains. Electrolysis has been shown to increase growth of the deep-sea cold-water coral Lophelia pertusa [25] and grow back eroded beaches [26]. Electric reefs could be grown on the deep-sea floor using batteries and fuel cells maintained by Remotely Operated Vehicles (ROVs) to repair deep cold-water reefs destroyed by trawlers, fossil fuel extraction, and deep-sea mining. Electrified enclosures in shallow atoll lagoons create intensely productive hatcheries and nurseries for juvenile sea cucumbers http://www.pacinternational.org/. Electric reefs therefore can be adapted to provide habitats from the surface of the ocean to the bottom, not only in shallow water.

## 6. Reef-wave interactions in time and space

The study of wave interactions with reefs languished because oceanographic research has been almost entirely funded by the military, which focused only on aspects related to military needs, such as whether waves would tip over military landing craft during invasions or detecting submarines [27], leaving fundamental coral reef research systematically unfunded. Research focused on large-scale oceanographic circulation around reefs [28] rather than understanding internal flows and processes, and what controls their variability. Because of the extreme complexity of reef structures, studies of wave-reef interactions can rarely be generalized beyond the reef that was measured. One focus has been to model impacts of extreme storm events that break corals to determine minimum forces to break corals [29]; however, corals vary enormously in morphology and internal strength, so the results rarely apply to more than one species.

Physical drag forces from water flow can be calculated from the Universal Drag Equation, $F_D = \frac{1}{2}C_D d A U^2$, where $F_D$ is the drag force on the object, $C_D$ is the drag coefficient, d is the density of the water, A is the cross-sectional area of the object perpendicular to the flow, and U is the velocity [30]. The Drag Coefficient represents the fraction of flow toward the structure that is blocked, is extremely sensitive to shape, surface roughness, and turbulence, and can range for very low for wires oriented along the flow vector to very large for flat objects perpendicular to it. The equation applies for simple geometrical objects with known drag coefficients, which few corals have, and steady flow, something that rarely happens in rapidly fluctuating coral reefs.

After the 2004 Asian Tsunami, Thomas Sarkisian, VP of the Global Coral Reef Alliance, and I studied his videos of coral reefs in the Similan Islands, Thailand, 1 week before and 1 week after the tsunami. In some patches, there was total coral destruction, mingled with patches with no damage at all even to fragile branching species, interspersed with areas of intermediate damage. The worst damage occurred where waves ran up the shore and flowed back full of rocks, soils, trees, wrecked buildings, and bodies, especially where wave energy focused on convex surfaces such as headlands and reef promontories (buttresses). Concave areas between promontories were largely untouched, indicating incredible temporal and spatial variability of wave breaking forces in even the most severe events and the difficulty of predicting their impacts theoretically.

Despite these limitations, important work has been done applying these principles to flow in the spaces between coral branches [31] finding that flow inside densely branched colonies can be nearly stagnant. On the other hand, spiny coral skeletons induce micro-scale turbulent vortex flows that greatly reduce the thickness of stagnant surface boundary layers and increase flow of oxygen and nutrients, stimulated by organized patterns of directed surface flows by ciliary pumping on the coral epithelia [1]. Measurements show internal pores in reefs are an important sink for bacteria and source of nutrients [32], but this obviously varies enormously in time and space depending on the size of the pores, their surface area, degree of interconnectedness, and throughflow.

Flow measurements on reefs show even simple reef structures with low-living coral cover greatly increase turbulent drag forces on reefs [33], but these can clearly be much higher in well-developed reefs with high-living coral cover [3]. A major focus has been to measure wave energy dissipation in waves crossing coral reefs. Due to the great expense of arrays of long-term recording current meters, this has only been done comprehensively by Storlazzi and colleagues on the South Shore of Molokai, where the reef has been shown to absorb up to 97% of the incident wave energy [34]. Clearly, these results are highly sensitive to overall reef profile, live coral cover, and kinds of corals, so cannot be easily generalized [35–37]. However, the lessons learned have been applied to estimate costs of reef decline and benefits of reef restoration for beach tourism, diving, and fishing in Florida, Puerto Rico, and Hawaii [38–40].

Without real field data that reflect the unique topography, wave climate, and history of each reef, the results are difficult to generalize, but it is clear coral reefs provide vastly superior environmental services at greatly reduced cost compared with concrete and rock seawalls. The models also show what we have long learned by practice in the field and are intuitively obvious that the greatest benefits for shore protection and natural beach regeneration come from reefs as high and close to shore as possible, with the highest porosity, permeability, and water throughflow. Electric framework reefs can be made as complex and as tall as needed, as long as they are physically stable. They have much lower mass, much larger negative space internal volumes, much higher interconnectedness resulting in high porosity and permeability, much higher internal surface area exposed to flow, and therefore biological activity than any other kind of artificial reef material, so they quickly achieve high biomass and biodiversity, often exceeding that of the former reef at the site that was restored. As of now, there are no flow performance measurements that allow these physical and biological benefits of regeneration to be quantified, but they are clearly visible in **Figures 3–5**.

During the 1947 Bikini Atoll Scientific Resurvey, physical oceanographer Walter Munk realized that the outer reef formed a regular series of growing coral reefs at right angles to the reef crest, oriented outward into the waves, separated by sand channels, and called this spur and groove morphology [41]. The constant pressure of waves on the reef crest resulted in water flow orientation into distinct cells, with broad incoming flow separated by narrow bands of much more intense return flow capable of eroding the reef, similar to riptide currents on beaches. The concentrated outflow eroded grooves into the reef rock, filled with rounded and abraded boulders of corals that had been smashed off the reef, rolling around in the channels, and living corals were rare on channel walls. Spur and groove formations were only found on sides of the atoll most exposed to prevailing winds and currents, and were absent on leeward slopes.

Thomas F. Goreau, research diver on the 1947 Bikini Atoll Scientific Resurvey, working with Roger Revelle measuring ocean and reef pH, $O_2$, $CO_2$, and alkalinity, built rebreathers to dive on coral reefs in Jamaica and found that these reefs also showed areas of coral growth oriented into the waves separated by sand channels, but their origin was entirely different, and they were caused by the rapid upward growth of corals, with sand canyons up to 30 meters deep between them, allowing excess limestone sediments produced by the reef to drain out into deep water. Unlike Bikini Atoll spur and groove morphology, with clearly erosional channels carved out of rock, Jamaican reef channels were depositional and completely covered with living corals, so he named it buttress and canyon morphology to highlight the fundamental ontogenetic difference with spur and groove reef structures [1, 42].

Unfortunately, constructional reef channels have largely continued to be called spur and groove by coral researchers, thereby confusing their origin. The two represent entirely different phases of coral reef growth and wave interaction. During rapid vertical reef growth, sand channels allow flushing of wave-driven water into the reef and the much larger volume of sediments produced by the reef to drain out into deep water. If the reef is shallow behind the reef crest, or fringing land, sand will accumulate in the lee of the reef, reducing outward flushing of sediment into deep water until the reef reaches sea level, and no further vertical growth can take place. At this point, the reef fills in and becomes a solid wall, greatly increasing the force of erosion on the forereef, as waves are now reflected since they can no longer flow over and through the reef. Waves reflected from a vertical wall receive a horizontal force that is twice the energy of the wave, because the wave reverses direction but does not lose strength in perfect reflection [42].

Reefs go through an aging cycle, starting from porous permeable rapidly growing structures with high throughflow, transforming into mature impermeable structures receiving the full force of the waves on the outer perimeter. Few measurements of reef physical, chemical, or biological performance take maturity of the reef structure into account, and this varies so much between reefs that measurements of reef flow in one reef will not be representative of another reef with different morphology, history, or wave climate, or even the same reef at a different stage in its life history. As a reef matures, the volume of internal negative spaces, their interconnections, and their throughflow, along with the physical, chemical, and biological processes dependent on them, are greatly reduced. As they evolve, reefs alter internal permeability through space filling growth of organisms in crevices, and through active formation of new holes by bioerosion of boring sponges, worms, sea urchins, clams, algae, and bacteria that pump water through the holes to meet their food and waste needs [1]. Finally, there are submarine lithification and cementation effects that depend on local factors. In high, wet, limestone islands like in Jamaica, the flow of groundwater percolating through the reef cements the entire reef framework into solid limestone rock, filling voids so thoroughly that the rock requires dynamite to excavate [43]. In sharp contrast, atoll reefs around low islands that lack groundwater flow, such as the Maldives, have essentially no internal reef cementation, and the entire framework consists of loosely attached corals that can easily be pulled apart from the reef by hand. On the other hand, some younger atolls, such as in the Tuamoto Archipelago, have internal hydrothermal circulation generated by residual heat from the buried volcanic formations around which the atoll grew as the volcano subsided [4], which draws deep ocean waters into the base of the atoll and causes internal cementation of sediments [44].

## 7. Electromagnetic fields and energy

The electrolysis process depends on flow of electrical current through the water and is therefore clearly affected by water flow patterns. Even though it has been long known that marine larvae are electrically polarized and migrate to the negative terminal in a DC electrical field, little work seems to have been done on electrotaxis of marine larvae since the 1970s although the method is used by scientists to collect fish eggs, fish larvae, and zooplankton [45]. Sea water electrolysis greatly enhances settlement of sessile marine organisms, and motile organisms are attracted in large numbers, especially juvenile fishes, as well as zooplankton at night [16, 46]. The electrolysis process appears to create biophysical conditions that increase biochemical energy production in marine organisms [15, 16]. The benefits are seen in increased growth, settlement, survival, and resistance to extreme stresses in all taxonomic groups, both animals and plants. The last universal common ancestor evolved the enzymes to make ATP and NAD used by all living organisms billions of years ago, perhaps using natural direct electrical currents in the sea caused by water movement, possibly in hot volcanic sea floor springs.

Since water is an electrical conductor, flowing through the earth's magnetic field as it is stirred by surface and internal waves, water currents, refraction, diffraction, absorption, reflection, etc., both electric and magnetic fields are constantly being generated. The fields generated are highest where there are strong water currents, and where waves break on shores or are focused on reef promontories, and may also be generated by fine-scale flow interactions with coral surface morphology. The electric and magnetic fields generated by breaking waves extend high into the atmosphere where they can be measured by satellites and deep into the ocean (Dommermuth, 2022, personal communication). Breaking waves cause an inverse energy cascade from small to large-scale motion that Dommermuth calls "The Ocean's Heartbeat," which drives the ocean's Langmuir circulation and stirs both the ocean and the lower atmosphere [47]. These natural fields can generate electrical currents high enough to exceed the sensitivity of sharks [48, 49]. Measurements of coastal turbulence show that the inverse energy cascade from small- to large-scale motion is largely driven by wave reflection from coasts and propagates to the interior of the ocean, rather than generated by open ocean wave breaking [50], which is no surprise since breaking free waves lack physical reaction forces caused by wave reflection [42]. There appear to be no direct measurements of the electrical and magnetic fields generated in coastal ocean waters, apparently due its lack of military importance (too shallow for submarines to hide).

Electrical currents are the driving force for of all biological energy production, but almost nothing is known about their natural or stimulated effects on marine environmental health except for natural history observations (16). The amount of energy radiated by the induced electromagnetic fields is given by the Poynting Vector, E x B, where E is the induced electrical field vector, B is the induced magnetic field vector, and the x represents the vector cross-product operation [51]. We hypothesize here that natural wave generated electromagnetic fields have fundamental biological benefits that are further stimulated by actively impressed electrolytic fields. In addition to electromagnetic fields developed in the water, eddy currents should also be induced by wave turbulence in the conductive steel frame itself. Direct measurements of induced low-frequency electrical and magnetic fields in the water, whose ultralow-frequency electromagnetic energy spectrum is unknown due to lack of direct measurements, will provide fundamental insights into these physical processes, their biological consequences, and how to manage them most effectively to regenerate marine ecosystems for climate change adaptation.

## 8. Conclusions: electric reefs for climate change adaptation

Reef ecosystems, and their invaluable and irreplaceable ecosystem services, are collapsing worldwide. Electric reef technology offers the first real hope of not only slowing the decline, but reversing it, and of amplifying the structural benefits beyond what could happen naturally, in site-specific problem-solving strategies. Corals are sick to death of humans, and electrolysis gives them an energy recharge to survive severe stress. Electrolysis technology will be an irreplaceable tool for adapting to climate change that already exceeds anything in human experience. Electric Reef Technology provides the first hope of reversing deterioration, regenerating ecosystems on large scales, increasing ecosystem service benefits even beyond those of natural reefs, and in locations where natural reefs cannot grow. More basic research is needed to understand the biophysical and biochemical mechanisms behind the benefits seen, and optimize them for rapid large-scale climate change adaptation and marine ecosystem regeneration.

Electric reefs mimic all functions of real coral reefs, but are more versatile because they can be rapidly built at places where thousands of years would be needed for natural growth under the best conditions. Electric reefs can be built in zones that are dead due to extreme conditions, severe pollution, even dead zones. Electric reefs grow back severely eroded beaches at record rates [21, 26] at a small fraction of the cost of seawalls and breakwaters that increase erosion in front of them, and with vastly greater benefits than concrete and rock structures by restoring marine ecosystem services and fisheries habitats [18]. They are the only method known to protect corals from dying during extreme high temperature events, which are becoming more frequent and severe [2].

Electric reefs grow solid rock frameworks of any form upward at up to 2 centimeters per year or more, 2–3 times harder than concrete [14]. They provide the only shore protection that grows to match sea level rise and provide adaptive protection that will never be overtopped by sea level rise, like seawalls will, only by extreme wave events. Electrolysis provides superior, lower-cost alternatives to protect coasts and regenerate collapsing fisheries to create sustainable Blue Economies and Blue Carbon sinks than any known alternative.

## Acknowledgements

I thank Wolf Hilbertz for 20 years of working together, for his lifetime of inspiration, and the many people who worked with us for decades around the world to develop his concepts. I thank Euen Jae Im and Lucy Wells for photographs, and Thomas Sarkisian, Douglas Dommermuth, and Nathalie Ledee for insightful discussions. This chapter is dedicated to Thomas F. Goreau, the first diving coral reef scientist, who discovered the incredible diversity of reef structure and function from shallow to deep, and was first to warn about human threats to them.

## Conflicts of interest

The author has worked on applications of the Biorock seawater electrolysis process with the late Wolf Hilbertz since the 1980s and coined the phrase "Biorock" as the generally accepted synonym for electrolytic marine materials previously called "Seacrete," "Seament," "Mineral Accretion," or "Electric Reefs" in order to emphasize its growing and auto-repairing nature, similar to corals. He has never made any money from this work; instead, it consumed all his very limited resources.

## Web links

https://www.globalcoral.org/biorock-brings-corals-back-in-ambon/
https://blueregeneration.com/#.

https://www.globalcoral.org/biorock-electric-coral-reefs-survive-severe-hurricanes-little-no-damage/
https://www.youtube.com/user/TheBiorockChannel
http://www.pacinternational.org/
https://en.wikipedia.org/wiki/Extra-low_voltage

## Author details

Thomas J. F. Goreau
Global Coral Reef Alliance, Cambridge, MA, USA

*Address all correspondence to: goreau@globalcoral.org

## References


[1] Goreau TF, Goreau NI, Goreau TJ. Corals and coral reefs. Scientific American. 1979;**241**:124-136

[2] Goreau TJF, Hayes RL. Global warming triggers coral reef bleaching tipping point. Ambio. 2021;**50**:1137-1140. DOI: 10.1007/s13280-021-01512-2

[3] Monismith SG. Hydrodynamics of coral reefs, Ann. Rev. Fluid Mechanics. 2007;**39**:37-55

[4] Darwin C. The Structure and Distribution of Coral Reefs. London: Smith; 1842

[5] Wallace AR. The Malay Archipelago. London: Macmillan; 1869

[6] Rumpf GE, Beekman EM. The Ambonese Curiosity Cabinet - Georgius Everhardus Rumphius. New Haven: Yale University Press; 1999

[7] Costanza R, de Groot R, Sutton P, van der Ploeg S, Anderson SJ, Kubiszewski I, et al. Changes in the global value of ecosystem services. Global Environmental Change. 2014;**26**:152-158

[8] Ferrario F, Beck MW, Storlazzi CD, Micheli F, Shepard CC, Airoldi L. The effectiveness of coral reefs for coastal hazard risk reduction and adaptation. Nature Communications. 2014;**5**:1-9

[9] Vousdoukas MI, Ranasinghe R, Mentaschi L, Plomaritis TA, Athanasiou P, Luijendijk A, et al. Sandy coastlines under threat of erosion. Nature Climate Change. 2020;**10**:260-263

[10] Hilbertz W, Goreau TJ. Third generation artificial reefs. Ocean Realm. 1998;**12**(3):45-48

[11] Foo SA, Asner GP. Sea surface temperature in coral reef restoration outcomes. Environmental Research Letters. 2020;**15**:074045

[12] Hilbertz WH. Electrodeposition of minerals in sea water: Experiments and applications. IEEE Journal of Oceanic Engineering. 1979;**4**(3):94-113

[13] Goreau TJ, Trench RK, editors. Innovative Technologies for Marine Ecosystem Restoration. Boca Raton: CRC Press; 2012

[14] Goreau TJ. Marine electrolysis for building materials and environmental restoration. In: Kleperis J, Linkov V, editors. Electrolysis. Rijeka, Croatia: InTech Publishing; 2012. pp. 273-290

[15] Goreau TJ. Electrical stimulation greatly increases settlement, growth, survival, and stress resistance of marine organisms. Natural Resources. 2014;**5**:527-537. DOI: 10.4236/nr.2014.510048

[16] Goreau TJF. Coral reef electrotherapy: Field observations. In: Seveso D, Downs C, Bhagooli R, editors. Cellular Stress Response and Physiological Adaptations of Corals Subjected to Environmental Stressors and Pollutants. Frontiers in Marine Science. Vol. 9. Lausanne, Switzerland: Frontiers Media S.A.; 2022. p. 805113. DOI: 10.3389/fmars.2022.805113

[17] Wells L, Perez F, Hibbert M, Clervaux L, Johnson J, Goreau T. Effect of severe hurricanes on biorock coral reef restoration projects in grand Turk, Turks and Caicos Islands. Revista Biologia Tropical. 2010;**58**:141-149

[18] Goreau TJ, Hilbertz W. Marine ecosystem restoration: Costs and benefits

for coral reefs. World Resource Review. 2005;**17**(3):375-409

[19] Galili E, Benjamin J, Eshed V, Rosen B, McCarthy J, Horwitz LK. A submerged 7000-year-old village and seawall demonstrate earliest known coastal defence against sea-level rise. PLoS One. 2019;**14**(12):e0222560

[20] Veyhl C, Belova IV, Murch GE, Öchsner A, Fiedler T. On the mesh dependence of non-linear mechanical finite element analysis. Finite Elements in Analysis and Design. 2010;**46**(5):371-378

[21] Goreau TJF, Prong P. Biorock reefs grow back severely eroded beaches in months. p. 243-263. Wahl T, Nilsen JEO, Haigh I, Brown S. Coastal Sea levels, impacts, and adaptation. Journal of Marine Science and Engineering. 2017;**5**(4):48. DOI:10.3390/jmse5040048

[22] Goreau TJ. Reef technology to rescue Venice. Nature. 2010;**468**:377

[23] Shaw R, Goreau TJ. Smart port breakwater design and construction. In: PORTS 2016. New Orleans: Proceedings of the American Society of Civil Engineers (ASCE)/Coasts, Oceans, Ports, and Rivers Institute (COPRI); 2016. pp. 863-872

[24] Goreau TJF. Biorock technology: A novel tool for large-scale whole-ecosystem sustainable mariculture using direct biophysical stimulation of marine organism's biochemical energy metabolism. In: Proceedings of the International Summit on Fisheries and Aquaculture. Amsterdam: Scientific Federation; 2018. p. 24

[25] Stromberg S, Lundalv T, Goreau T. Suitability of mineral accretion as a rehabilitation method for cold-water reefs. Journal of Experimental Marine Biology and Ecology. 2010;**395**:153-161

[26] Goreau TJ, Hilbertz W, Azeez Hakeem A, Sarkisian T, Gutzeit F, & Spenhoff A. Restoring reefs to grow back beaches and protect coasts from erosion and global sea level rise. in Goreau TJ & Trench RK, Editors. Innovative Technologies for Marine Ecosystem Restoration. 2012: Boca Raton, CRC Press

[27] Oreskes N. Science on a Mission: How Military Funding Shaped What we Do and Don't Know about the Oceans. Chicago: University of Chicago Press; 2020

[28] Wolanski E, Pickard GL. Physical Oceanographic Processes of the Great Barrier Reef. Boca Raton: CRC Press; 2018

[29] Madin JS, Connolly SR. Ecological consequences of major hydrodynamic disturbances on coral reefs. Nature. 2006;**444**:477-480

[30] Wiegel RL. Oceanographical Engineering. Minneola, New York: Dover; 1992

[31] Chamberlain JA, Graus RR. Water flow and hydromechanical adaptations of branched reef corals. Bulletin of Marine Science. 1975;**25**:112-125

[32] Van Duyl FC, Scheffers SR, Thomas FIM, Driscoll M. The effect of water exchange on bacterioplankton depletion and inorganic nutrient dynamics in coral reef cavities. Coral Reefs. 2006;**25**:23-36

[33] Riedenbach MA, Monismith SG, Kosseff J, Yahel G, Genin A. Boundary layer turbulence and flow structure over a fringing coral reef. Limnology and Oceanography. 2006;**51**:1956-1968

[34] Storlazzi CD, Ogston AS, Bothner MH, Field ME, Presto MK. Wave- and tidally-driven flow across a

fringing coral reef: Southern Molokai, Hawaii. Continental Shelf Research. 2004;**24**:1397-1419

[35] Storlazzi CD, Jaffe BE. The relative contribution of processes driving variability in flow, shear, and turbidity over a fringing coral reef: West Maui, Hawaii. Estuarine, Coastal, and Shelf Science. 2008;**77**:549-564

[36] Rogers JS, Monismith SG, Feddersen F, Storlazzi CD. Hydrodynamics of spur and groove formations on a coral reef. Journal of Geophysical Research: Oceans. 2013;**118**:3059-3073

[37] Rogers JS, Maticka SA, Chirayath V, Woodson CB, Alonso JJ, Monismith SG. Connecting flow over complex terrain to hydrodynamic roughness on a coral reef. Journal of Physical Oceanography. 2018;**48**:1567-1587

[38] Storlazzi CD, Reguero BG, Viehman TS, Cumming KA, Cole AD, Shope JA, et al. Rigorously Valuing the Impact of Hurricanes Irma and Maria on Coastal Hazard Risks in Florida and Puerto Rico. Washington DC: U.S. Geological Survey Open-File Report 2021-1056; 2021. p. 29. DOI: 10.3133/ofr20211056

[39] Storlazzi CD, Reguero BG, Cumming KA, Cole AD, Shope JA, Gaido LC, et al. Rigorously Valuing the Coastal Hazard Risks Reduction Provided by Potential Coral Reef Restoration in Florida and Puerto Rico. Washington DC: U.S. Geological Survey Open-File Report 2021-1054; 2021. p. 35. DOI: 10.3133/ofr20211054

[40] Storlazzi CD, Reguero BG, Yates KK, Cumming KA, Cole AD, Shope JA, et al. Rigorously Valuing the Impact of Projected Coral Reef Degradation on Coastal Hazard Risk in Florida. Washington DC: U.S. Geological Survey Open-File Report 2021-1055; 2021. p. 27. DOI: 10.3133/ofr20211055

[41] Munk WH, Sargent MC. Adjustment of bikini atoll to wave. EOS, Transactions of the American Geophysical Union. 1948;**29**:855-8605. DOI: 10.1029/TR029i006p00855

[42] Goreau TF. A Study of the Biology and Histochemistry of Corals [thesis]. United States: Yale University; 1956

[43] Land LS, Goreau TF. Submarine lithification of Jamaican reefs. Journal of Sedimentary Research. 1970;**40**:457-462

[44] Rougerie F. Nature and functioning of the atolls of tuamoto archipelago. French polynesia. Oceanologia Acta. 1995;**18**:61-78

[45] Pearl R. Recent work in electrotaxis. The American naturalist. 1900;**34**:977-979 Oceanologica Acta. 18:61-78

[46] Kelso WE, Kaller MD, Rutherford DA. Collecting, processing, and identification of fish eggs and larvae and zooplankton. In: Fisheries Techniques. 3rd ed. Bethesda: American Fisheries Society; 2012. pp. 363-452

[47] Dommermuth DG. The ocean's heartbeat: An inverse energy cascade that mixes the lower atmosphere and the upper ocean. In: 34th Symposium on Naval Hydrodynamics. Washington DC: Research Gate Preprint; 2022. Available from: https://www.researchgate.net/publication/360983504

[48] Kalmijn AJ. The electric sense of sharks and rays. Journal of Experimental Biology. 1971;**55**:371-383

[49] Uchoa MP, O'Connell CP, Goreau TJ. The effects of biorock-associated electric fields on the Caribbean reef shark

(*Carcharhinus perezi*) and the bull shark (*Carcharhinus leucas*). Animal Biology. 2017;**67**:191-208

[50] Elgar S, Raubenheimer B. Field evidence of inverse energy cascades in the surfzone. Journal of Physical Oceanography. 2020;**50**(8):2315-2321. DOI: 10.1175/JPO-D-0327.1

[51] Susskind L, Friedman A. Special Relativity and Classical Field Theory. New York: Basic Books; 2017

Chapter 5

# Coral Reef Deterioration and Livelihoods of Coastal Communities: An Economics Perspective

*Pasita Chaijaroen*


## Abstract

This chapter focuses on how coral reef deterioration impacts the well-being of the affected coastal communities. In a macro-level perspective, the chapter discusses 1) how coral reefs socioeconomically benefit humans and 2) how coral reef deterioration impacts humans. This part of the discussion sheds light on several adverse effects of coral reef degradation ranging from reductions in food availability and income to losses of jobs and built capital. In a micro-level perspective, the socioeconomic impact of coral reef degradation on vulnerable coastal communities is illustrated through a case study of the 1998 coral bleaching in Indonesia. The households affected by this bleaching event experienced a large income shock, which translated into reduced protein consumption and impeded child development. The chapter then concludes with two broad policy recommendations drawn from both the macro- and micro-level discussions. First, reef conservation and restoration are economically viable investments due to the immense socioeconomic values of coral reefs. Second, policy interventions are required to mitigate impacts of coral reef degradation on the vulnerable groups and to facilitate the adaptation process.


**Keywords:** human-coral interactions, coral reef degradation, coral bleaching, Indonesia, economic values of coral reefs

## 1. Introduction

The oceans cover more than 70% of our planet [1] and provide immense socioeconomic benefits to humans. For example, the oceans provide livelihoods to more than three million people worldwide [2]. In addition, 350 million jobs around the world are related to the oceans, and one billion people in developing countries rely on fish as their main source of protein [1]. These benefits on jobs and income are enjoyed more by developing countries rather than by the Organization for Economic Cooperation and Development (OECD) countries [2], suggesting the important role of marine resources in global socioeconomic development.

Coral reefs benefit the marine ecosystems in various ways—many of which translate into goods and services that sustain human livelihoods. **Table 1** exhibits that humans derive various goods from the coral reefs, from food products to raw materials and live fishes [3, 4]. Coral reefs also provide crucial services to both the marine ecosystems and humans. For example, coral reefs serve as habitats for juvenile fishes and help protect shorelines against waves and storms.

Coral reefs are closely tied to the livelihoods of millions of people worldwide. Each year, almost $30 billion in net benefits are realized from three main functions of coral reefs: fisheries production, tourism, and shoreline protection [5]. When accumulated

| | | |
|---|---|---|
| Goods | Renewable resources | • Seafood products |
| | | • Raw materials/starters for medicines |
| | | • Raw materials (seaweed, materials for jewelry and decoration, etc.) |
| | | • Live fish and corals for aquarium |
| | Mining of reefs | • Coral blocks, rubble, and sand for building |
| | | • Raw materials for lime and cement |
| | | • Mineral oil and gas |
| Ecological services | Physical structure services | • Shoreline protection |
| | | • Buildup of land |
| | | • Promoting growth of mangroves and seagrass beds |
| | Biotic services | • Maintenance of habitats |
| | | • Maintenance of biodiversity and a genetic library |
| | | • Regulation of ecosystem processes and functions |
| | | • Biological maintenance of resilience |
| | | • Biological support of other ecosystems via â€œmobile linksâ€ |
| | | • Export of organic production and plankton to pelagic food webs |
| | Biogeochemical services | • Nitrogen fixation |
| | | • Carbon dioxide and calcium budget control |
| | | • Waste assimilation |
| | Information services | • Monitoring and pollution record |
| | | • Climate record and control |
| | Social and cultural services | • Recreation activities |
| | | • Esthetic values and artistic inspiration |
| | | • Sustaining the livelihood of communities |
| | | • Parts of cultural, religious, and spiritual values |

*Source: Adapted from [3, 4].*

**Table 1.**
*Good and services provided by coral reefs.*

over a 50-year period, the net benefits will total to almost $800 billion in net present value terms [5].

Unfortunately, coral reefs around the world have been suffering from various stressors in recent years. As of 2003, 27% of coral reefs were permanently lost [5]. 90% of all coral reefs might be lost by 2050 even when all the objectives of the Paris Agreement are accomplished [6]. Human threats account for the majority of coral reef degradation and include activities such as coastal development, sand and coral mining, overfishing and destructive fishing, and water-shed pollution [7]. Natural threats to coral reefs include storms [8], ocean acidification, and coral bleaching events [7]; these natural threats are often exacerbated by human activities.

Human livelihoods are closely tied to the oceans; therefore, coral reef degradation could impose significant socioeconomic threats on economies and communities around the world. Section 2 will discuss economic benefits provided by coral reefs in the three main socioeconomic activities: fisheries, tourism, and shoreline protection. The section will also paint a picture of socioeconomic losses due to coral reef degradation in these sectors.

While the socioeconomic losses from coral reef degradation will be experienced worldwide, the most vulnerable group is small island nations and developing countries near the oceans. Some low-income countries and small island states derive more than 20% of their GDP from tourism and other ocean-related sectors compared to the less than 2% in high-income countries [2]. These vulnerable states usually have with few other economic and livelihood alternatives. Nine countries have been identified by [7] as most vulnerable to coral reef degradation: Haiti, Grenada, Philippines, Comoros, Vanuatu, Tanzania, Kiribati, Fiji, and Indonesia. These countries rely on reefs significantly with limited adaptation capability, but their reefs are facing high levels of threats. These aspects will be explored in Section 3 using a case study of the 1998 mass coral bleaching in Indonesia.

While some coral reefs were deemed permanently lost [5], some can still be replenished and restored. Coral reef conservation and restoration are usually economically viable because their benefits dwarf their costs by several times. Section 4 will discuss this policy aspect in detail. In addition to conservation and restoration, adverse effects of coral reef degradation call for policy interventions in terms of mitigation and adaptation. These policy implications will also be elaborated in Section 4.

## 2. Key socioeconomic benefits from coral reefs and consequences of reef degradation

Humans rely on coral reefs for various goods and services (see **Table 1** for detail). Each year, almost $30 billion in net benefits are realized from coral reefs[1] [5]. Of the many goods and services that human derive from coral reefs, three key ones have received significant attention in the economic literature: fisheries, tourism, and shoreline protection (see **Table 2**). Each of these functions will be discussed in detail as follows.

---

[1] Net benefits are defined as total benefits minus total costs. When accumulated into a net present value term over a 50-year period, these net benefits total to almost $800 billion, assuming a discount rate of 3%.

| Goods/services | Potential net benefits |
|---|---|
| Tourism/recreation | 9.6 |
| Coastal protection | 9.0 |
| Fisheries | 5.7 |
| Biodiversity value | 5.5 |
| Total | 29.8 |
| *Source: [5].* | |

**Table 2.**
*Annual potential net benefits of coral reefs in USD billions.*

## 2.1 Fisheries

Coral reefs are known for their significant contribution to the abundance of fish, from those that live on the reefs to those that rely on the reefs ecologically. For example, coral reefs serve as grounds for spawning, nursery, breeding, and feeding for many species [3]. Coral reefs also export organic matters and different types of plankton and, hence, support the pelagic food web [3]. Coral reefs, therefore, provide immense economic values to fisheries. For instance, in 2017, commercial fisheries in the Coral Triangle area and the Mesoamerican Reef provided up to $5.85 billions and $480 millions in total economic returns, respectively [9].[2]

Income is not the only benefit that humans derive from fisheries; fish is a significant source of protein for humans. 3.3 billion people around the world derive 20% or more of their animal protein intake from fish [10]. Fish accounts for at least 50% of animal protein consumption in many developing nations, e.g., Bangladesh, Ghana, and many small island developing states [10]. In particular, many developing island nations depend heavily on coral reefs as a food source; island populations usually have limited options when it comes to protein sources [9].

In addition to providing food, the fisheries sector also provides around 39 million people around the world with jobs [10] and, consequently, supports their and their family's livelihood. Similar to the reliance for consumption, developing and small island countries are more likely to depend on the oceans for jobs than other countries. Most of the people working in fisheries and aquaculture are in developing countries and in small-scale, artisanal fisheries [10].

Coral reef deterioration can translate into adverse effects in the fisheries sector. While coral reef degradation can improve reef fisheries productivity in the short term when benthic turf and invertebrates increase and reef structures are still intact [11], the long-term effects are likely to be negative [11, 12]. Abundance of many species targeted by reef fisheries will eventually decline because of changes in habitat structures as well as food sources [12].

Given the world's significant reliance on fisheries, coral reef deterioration can have devastating effects on income, jobs, and consumption of millions of people around the world. Small-scale and artisanal fisheries have been identified as more vulnerable to

[2] The total economic returns consist of the direct economic impact from fisheries income and the indirect economic impact from the fisheries sector's spending on other sectors' goods and services. This report focuses only on commercial fisheries; therefore, the reported returns are underestimated.

changes in fish stocks and fishing conditions than commercial fisheries because they are more constrained, for instance, by limited mobility, capital, and access to credits [13, 14]. Cultural values, limited access to international markets, and a lack of alternative livelihoods also reduces artisanal fisheries' capacity to adapt to these changes [14–16].

### 2.2 Recreation and tourism

Coastal tourism constitutes a large part of the travel and tourism industry, one of the largest industries in the world. Coral reefs are the backbone of coastal tourism, from serving as dive sites to providing by-product attractions such as white-sand beaches, serene seas, and beautiful waters. Each year, over 350 million people visit coral reef coasts in 102 out of the 117 countries and territories with coral reefs around the world [6].

The tourism value of coral reefs is estimated to be $37.8 billion per year. This value, however, is derived from only about 30% of the world's coral reefs because the rest of them are too remote [6]. For many countries, coral reefs are vital to their income, economy, and livelihoods. For example, 26 countries/territories earn more than a quarter of their GDP from tourism. In addition, many major coral-reef destinations are in developing small island nations which have few other alternative livelihoods [6].

Coral reef degradation, therefore, could have tremendous effects on many economies around the world. Coral reef degradation could lead to declines in the number of tourists [17] and the number of dives Ceasar et al. [18], as well as destination changes [19, 20] and reduced willingness to pay for coral reefs [17, 19]. While parts of the economic losses are lost income to businesses and workers in the tourism sector, a significant portion of these losses are tourists' loss of welfare. Many tourists are aware of coral reef degradation, though the degree varies by destinations [17, 20, 21]. This awareness is usally associated with reduced tourist satisfaction [17, 19, 22] resulting in a lower willingness to pay for coral reefs and decreased tourist welfare [17, 19].

Past research has shown that total economic losses in tourism from coral bleaching range from several thousands to millions of US dollars per site event [17, 18, 23]. For example, the 1998 coral bleaching event resulted in approximately $350,000 loss in total welfare per year in Palau [17]. The coral bleaching event in 2010 was associated with an economic value loss of $50–80 millions in Thailand, Indonesia, and Malaysia [23].

### 2.3 Shoreline protection

More and more of the world's population are living near coastlines. Almost 2.4 billion people live within 100 km of the coast; people living in coastal communities represent 37 of the world's population as of 2017 [24]. In addition to housing a significant proportion of the world's population, the coastlines around the world are also filled with built infrastructure and properties. Damages to coastal areas, therefore, extend to not only humans but also the land and built capital.

Risks across the shoreline, including potential damages from flood and storm, have been increasing. In the first decade of the 2000s, 4.3% of the world's GDP are susceptible to tropical cyclones, a significant increase from the 3.6% in the 1970s [25]. Risks of economic losses from cyclones increased in all regions, with the average annual GDP exposed to cyclones tripled from 525.7 billion in the 1970's to 1576.5 billion in the first decade of the 2000s. While the average global GDP exposed to floods is smaller at 100 billion in the first decade of the 2000s, this exposure also tripled from that in the 1970s [25].

Coral reefs have been found to play a significant role in protecting shorelines from land erosion and damages from floods and storms. This protection extends over 150,000 km of coastlines in more than 100 countries around the world [7]. According to a meta-analysis of over 200 studies, coral reefs, on average, reduce wave energy by 97% and, therefore, protect shorelines against winds and storms. Most of this protection (86%) is derived from reef crests [8]. Over 100 million people are expected to benefit from coral reefs' shoreline protection and, hence, the reduced risks of flooding and storm damages. Specifically, up to 100 million people are living below 10-m elevation and within 10 km of a reef. Extending this range a little bit further to within 50 km of reefs, there are about 197 million people who live below 10-m elevation. **Table 3** exhibits the 15 countries with the highest number of people living in this latter range.

When quantified into dollar amounts, shoreline protection benefits could dwarf economic benefits from fisheries, aquaculture, and tourism [27]. A recent global study [26] finds that coral reefs reduce expected damages from storms by more than $4 billion per year. With a 1-m change in reef profile (i.e., the without reef scenario), the storm damages could have more than doubled (118%). In addition, flooding of land would have increased by 69% and affected 81% more people each year in the without

| No. | Number of people | | Annual averted damages | | Annual averted damages/GDP | |
|---|---|---|---|---|---|---|
| 1 | Indonesia | 41 | Indonesia | 639 | Cayman Islands | 0.98 |
| 2 | India | 36 | Philippines | 590 | Belize | 0.37 |
| 3 | Philippines | 23 | Malaysia | 452 | Grenada | 0.3 |
| 4 | China | 16 | Mexico | 452 | Cuba | 0.25 |
| 5 | Vietnam | 9 | Cuba | 401 | Bahamas | 0.16 |
| 6 | Brazil | 8 | Saudi Arabia | 138 | Jamaica | 0.14 |
| 7 | United States | 7 | Dominican Rep. | 96 | Philippines | 0.13 |
| 8 | Malaysia | 5 | United States | 94 | Antigua & Barbuda | 0.13 |
| 9 | Sri Lanka | 4 | Taiwan | 61 | Dominican Rep. | 0.11 |
| 10 | Taiwan | 3 | Jamaica | 46 | Malaysia | 0.09 |
| 11 | Singapore | 3 | Vietnam | 42 | Seychelles | 0.06 |
| 12 | Cuba | 3 | Myanmar | 33 | Turks & Caicos | 0.06 |
| 13 | Hong Kong | 2 | Thailand | 32 | Guadeloupe | 0.05 |
| 14 | Tanzania | 2 | Bahamas | 14 | Indonesia | 0.04 |
| 15 | Saudi Arabia | 2 | Belize | 9 | Solomon Islands | 0.04 |

*Sources: Adapted from [8, 26].*
Notes: *This table illustrates top 15 countries that benefit most from coral reef shoreline protection in terms of the number of people living below 10-m elevation and within 50 km of the coast (in million) [8]. The annual averted damages cover damages to built capital in USD millions per year. The annual averted damages relative to GDP show each country's damage size relative to its economy size [26].*

**Table 3.**
*Countries with largest benefits from coral reefs' shoreline protection in terms of the number of people living in coastal areas and the annual averted damages on coastal built capital.*

reef scenario. **Table 3** provides a list of countries that enjoy the highest annual flood protection benefits in this hypothetical situation. Most of these countries are developing; some are small island nations. In a more extreme case, flood damages from a 100-year storm event would have been $272 billion, a 91% increase from the scenario with the reefs. At a country level, this extreme-case benefit of reefs surpasses $1 billion for more than 10 countries around the world, most of which are developing countries. Regardless of whether the protection benefits are for annual floods or a 100-year storm event, small island nations in the Caribbean and the South Pacific see the highest benefits relative to their GDPs.

To sum up, this section provides a macro-level overview on how coral reef degradation can result in tremendous socioeconomic losses, ranging from reductions in food availability, income, and welfare to losses of jobs and built capital. Developing countries, including small island nations, have been identified as more vulnerable to coral reef degradation because their economies and livelihoods depend significantly on the oceans.

To illustrate how coral reef deterioration affects a vulnerable group, the next section will discuss micro-level socioeconomic impact through a case study in one of most vulnerable developing island nations. Oftentimes, an economic shock creates rippled effects far beyond its first-order effects. Coral reef deterioration is no exception. The case study below will also exhibit how coral reef deterioration affects not only income but also other socioeconomic aspects of the affected coastal communities.

## 3. Case study: the 1998 coral bleaching and the Indonesian fisheries sectors

The mass coral bleaching in 1998 was the most severe bleaching event up until its occurrence [28, 29]. High sea surface temperatures and bleaching spots were reported throughout the tropical zone worldwide [29, 30]. 16% of the world's corals were lost in this bleaching event [29]. In Indonesia, coral bleaching spots were reported in West Sumatra, Central Java, Bali and Lombok area, Southern Sulawesi, and Papua; with mortality rates up to 50% [30]. This bleaching event could potentially pose significant threats on the Indonesian fisheries sector, most of which were considered small scale [31][3] and, hence, particularly susceptible to the changes in ocean conditions and fish stocks.

This section presents how the severe episode of coral bleaching in 1998 affected the lives of Indonesian fisheries households—from their economic well-being to their health and other socioeconomic outcomes. The discussion in this section is based on findings from two studies. The first study [32] highlights how the bleaching affected income, labor outcomes, and consumption. The second study [33] demonstrates how the bleaching event impacted reproductive decisions as well as how the economic hardship from coral bleaching might have adverse effects on child development.

Both studies relied on household survey data from 1993, 1997, 2000, and 2007 in conjunction with coral bleaching data from 1998.[4] The affected households, namely

[3] More than half of the marine fishing boats in Indonesia were nonpowered in 2000 [31].

[4] Both studies merged the household survey data from the Indonesian Family Life Surveys [34–37] with reported coral bleaching spots from [30] and sea surface temperature anomalies published by the National Oceanic and Atmospheric Administration (NOAA).

fisheries households in the areas with reported coral bleaching, were compared against other households using regression analyses that controlled for a number of potential confounding factors.[5]

The first study shows that the 1998 bleaching event socioeconomically affected the households in three ways [32]. First, the affected households experienced a significant decline in income—the average of 27%—in 2000 but not in 2007.[6] Lower fishery yields were one plausible reason for this income shock. The lack of alternative livelihoods, at least in the short term, could have also contributed to this income shock. In 2000, the affected households could not increase their working hours or find new or secondary jobs potentially because fisheries skills might not be valuable in other jobs. Moreover, fishery workers usually have lower education than other workers. Evidence from 2007, however, shows more adaptation capacity. The affected households were then more likely to switch to a new industry and to increase their labor supply relative to other households.

Finally, the affected fisheries households consumed less protein in both 2000 and 2007, but other consumption was generally unaffected. Some consumption also increased in 2007 as income improved. The reduction in protein consumption came as no surprise because the affected households rely on the ocean for both their income and food sources. Nonetheless, this protein consumption shock raises concerns over nutrition intakes and other consequential outcomes, especially those related to child development.

In fact, economic shocks usually affect fertility decisions as well as various child development and later-in-life outcomes. Economic shocks may increase or decrease fertility [38–40]. In the case of the 1998 coral bleaching in Indonesia, the affected households were more likely to have children relative to other households in 2000 [33]. This rise in fertility was only a temporary shift of fertility timing rather than a permanent increase in the total number of children. One plausible reason is the temporary lowered opportunity costs of children due to limited job prospects in 2000.

Unfortunately, some evidence in 2000 points toward an increase in the likelihood of severe malnutrition among the children born to the affected households. By 2007, the affected households already enjoyed a significant improvement in income, but their children were still more likely to fail a grade in school despite their higher enrollment rate than the other children's. These findings are consistent with a large literature in economics in which shocks in the first 5 years of life usually affect children's health and schooling outcomes as well as have long-term consequences into adulthood [41–43].

What is striking about the results in [33] is that the affected households decided to temporarily increase their fertility, but their children's health and schooling outcomes were inferior to their peers'. This finding along with the other adverse effects of coral bleaching on income, labor market outcomes, and consumption sheds light on how coral reef degradation can affect many socioeconomic dimensions, even when the analyses were restricted to just one economic sector. These findings, together, call for comprehensive policy interventions in mitigation and adaptation.

---

[5] The control factors include household socioeconomic status prior to coral bleaching, time-invariant factors within each household and province, and time-varying factors that were experienced nationwide, e.g., the Asian economic crisis in 1998.

[6] The average 27% reduction in income is an estimate from the most preferred specification. Estimates from other specifications ranged between 27% and 46%.

While the findings discussed in this chapter may be applicable to similar coral reef degradation events in similar parts of the world, these findings will not extrapolate to all cases. Different countries usually have different economic activities, market structures, cultures, and norms. Coral reef degradation also differs in terms of spatial and temporal scales as well as severity. More work in this area is definitely needed to better understand micro-level linkages between coral reefs and humans.

## 4. Policy implications

The immense economic values of coral reefs discussed in Section 2 imply that coral reef conservation and restoration are economically sound investment. In addition, the case study in Indonesia in the previous section highlights the needs for policy makers to step in and help facilitate mitigation and adaption as coral reefs deteriorate. This section will broadly discuss these two types of policies in further detail.

### 4.1 Coral reef conservation and restoration

One concept related to the oceans that has been gaining traction in recent years is blue economy, a concept that integrates economic prosperity from marine resources with sustainability [6]. The world economy depends heavily on marine resources; yet, these resources are facing with many risks and challenges. Small island nations and many coastal countries then have to balance between growing the ocean economy and maintaining healthy ocean ecosystems [6]. This subsection will discuss the blue economy policy agenda related to coral reefs. It will first highlight the economic viability of coral reef conservation and restoration and then will elaborate on some major policy tools.

Maintaining healthy coral reefs has been proven to be an economically sound policy agenda. An analysis of economic returns [6] to different coral reef statuses suggests that bringing the reefs to a healthy status is worth the investment with returns up to 44 times. For instance, an additional $37 billion in economic value can be unlocked if the reefs in Indonesia are replenished back to the healthy status by 2030. This type of analyses implies that the private sector as well as governments and nongovernmental organizations (NGOs) should take parts in reef conservation and restoration, and that economic prosperity can conform with sustainability.

In addition, coral reef restoration makes economic sense because restoration is oftentimes the most cost effective policy available. For example, reef restoration is cost effective in protecting coastlines when compared to artificial defenses [8].

In addition to the estimated economic value above, coral reef conservation and restoration could provide significant societal benefits such as health benefits from water pollution abatement and carbon capture from coastal afforestation. The expansion of no-take zones, as another example, helps preserve fish stocks and biodiversity.

Owing to the large potential benefits, the restoration of coral reefs has gained salience in recent years. By 2013, nurseries across the Caribbean, Southeast Asia, and the Pacific and Indian Oceans raised more than 86 coral species and over 100,000 colonies [6].

For marine resource conservation, marine-protected areas (MPAs) is a commonly used policy tool. For coral reefs, studies have shown that a partial closure (20–40% of all reef areas in a location) could lead to net increases in fishery catches in remaining areas [6]. However, the current MPAs only cover about 4% of the world's ocean surface [6], while a 30–40% of coverage is needed for an effective protection[7] [44]. In

| | Mesoamerica reef | | | Coral triangle | | |
|---|---|---|---|---|---|---|
| | Costs | Benefits | ROI (times) | Costs | Benefits | ROI (times) |
| No-take zone | 30 | 1321 | 44.0 | 183 | 1618 | 8.8 |
| Vegetative filter strip | 5 | 24 | 4.8 | 61 | 206 | 3.4 |
| Afforestation | 7 | 25 | 3.6 | 212 | 453 | 2.1 |
| Constructed wetlands | 8 | 22 | 2.8 | 193 | 134 | 0.7 |

*Source: Compiled from [5].*
Notes: *This table illustrates the projected present values of costs and benefits along with returns on investments (ROI, in multiples) in four conservation initiatives in the Mesoamerica Reef and the Coral Triangle.*

**Table 4.**
*Returns to investments in conservation in the Mesoamerican reef and the coral Traingle area.*

terms of matching, the current MPAs coverage matches well with coral reefs most important in tourism, but the coverage of reefs that are important for shoreline protection is not as comprehensive [6].

Other policies options have also been proven to be economically sound. For example, four policy interventions have been evaluated for the Mesoamerican Reef and the Coral Traingle Area: 1) expansion of no-take MPAs, 2) implementation of vegetative filter strip to reduce erosion, 3) afforestation to reduce erosion, and 4) construction of wetlands to treat wastewater [6]. Results show positive and high returns on investments in all but one case (see **Table 4**).[8]

## 4.2 Policies to support mitigation and adaptation

While restoration and conservation could mitigate coral reef deterioration, some degradation is still expected, for example, due to climate change. As discussed in Section 3, coral reef degradation can adversely affect people whose livelihood depends on the ocean in many ways, ranging from income losses to lowered consumption and impeded child development. These negative socioeconomic impacts imply the needs for comprehensive policy interventions to 1) mitigate these impacts and 2) facilitate adaptation. This subsection outlines the policy implications drawn from the case study in Section 3 and briefly discusses how they might extrapolate to other cases.

The first step in policy intervention is to mitigate the adverse effects of coral reef degradation on income and consumption. Protein and nutrition supplements could provide a direct consumption support. Conditional cash transfers[9] can also be

[7] The meta-analysis [44] identifies six marine conservation objectives: 1) to protect biodiversity, 2) to ensure connectivity, 3) to avoid collapse, 4) to avoid adverse evolution, 5) to optimize fisheries value, and 6) to satisfy multiple stakeholders. Different studies call for different coverages to meet these objectives. The 30–40% coverage is the range in which many studies converge.

[8] **Table 4** presents the returns on investment in multiples, that is, how many times each initiative's benefits outweigh its costs. In all but one case, the returns on investments are greater than 2, suggesting that the benefits are more than twice the costs. The exception is the constructed wetlands in the Coral Triangle area, which have higher costs than benefits due to Indonesia's low access to wastewater treatment and high amount of untreated wastewater [6].

[9] Conditional cash transfers are welfare programs that condition the payments or transfers on receivers' actions such as enrolling in school or receiving a health checkup.

effective in alleviating the immediate income and consumption shocks while lessening further negative impacts, e.g., on child development. One reason is that consumption increased with income; therefore, a cash transfer could have boosted consumption. By conditioning the transfers on child-related outcomes, such as prenatal care and school enrollment, these transfers would also support child development. In fact, conditional cash transfers have been shown to improve preventive healthcare utilization, increase protein consumption, and decrease wasting and stunting in children both in Indonesia and elsewhere in the world [45–49].

Policies to mitigate labor and fertility outcomes and facilitate adaptation in a longer term should target both capacity building and labor market friction reduction. Specifically, capacity building is required in a number of areas. For example, in the case study in Section 3, the temporary increase in fertility followed by inferior children outcomes suggests that the affected households might have underestimated the effects of coral bleaching. Educating these households about the plausible long-term effects could have helped them make informed fertility decisions [33]. Capacity building in terms of skill acquisition could also facilitate labor adjustments such as adoption of new fishing gears and techniques, finding a part-time job, or switching to a new industry. Labor market frictions might have also contributed to the delay in labor adjustments after the 1998 coral bleaching in Indonesia. Policies that reduce search costs and facilitate information flows, e.g., job matching services, could also help expediting labor market adjustments [32].

While these policy recommendations are drawn from one specific coral bleaching event in Indonesia, they can be applicable to other bleaching events as well as coral reef degradation in general with some caveats. Many countries that are suffering or will suffer from coral reef degradation are developing countries and island states with many characteristics similar to Indonesia. The problems that these countries face are generally similar, ranging from low/moderate income to inadequate accesses to good education and healthcare. In this sense, the general policy recommendations outlined in this subsection will also apply to these countries. Nonetheless, developing countries usually differ in finer details. For example, one country might have a basic universal healthcare coverage, while others do not. Compulsory schooling is six years in some countries and nine years in others. For this reason, specific policy recommendations should be tailor-made to suit each country's coral reef prospects and socioeconomic environment.

## 5. Conclusion

Coral reef deterioration can impact humans in many ways. From a macro-level perspective, coral reef deterioration can pose significant threats on global food availability, income, job security, and built capital along the coastlines, among others. From a micro-level perspective, coral reef deterioration can affect several dimensions of human life. For example, a case study on the 1998 coral bleaching in Indonesia in Section 3 reveals the adverse effects of coral bleaching on household income, protein consumption, labor market outcomes, fertility decisions, and child development.

The adverse effects of coral reef deterioration call for policy intervention in two key dimensions. First, coral reef conservation and restoration are usually economically sound due to immense socioeconomic threats from coral reef deterioration. A number of conservation and restoration policy tools are discussed in Section 4.1. Second, policy interventions are required to mitigate the effects of coral reef degradation and to facilitate adaptation, as shown through the 1998 coral bleaching case study in Indonesia. These interventions were discussed at length in Section 4.2.

While this chapter covers many elements of the impact of coral reef deterioration on humans, further studies in various areas are still needed. Most of the current literature on human-coral interactions only covers certain functions of some coral reefs. Given the vast differences across different coral reefs and the coastal communities around them, further work is required to better understand these human-coral interactions in greater detail. For example, other functions of coral reefs apart from the main ones mentioned in this chapter should be studied, and future works should be done in many more countries and geographical areas. One of the main reasons for the scarce literature in this field is the lack of data sets that contain information on both coral reef status and the socioeconomic status of the local communities. Advancement in data technologies such as remote sensing and geographic information system (GIS) could bring out more detailed studies in the future. These studies will be crucial for designing tailor-made policies that match well with each local environment.

Finally, despite the clear evidence on the importance of reef conservation and restoration, policy implementation in the real world is still lagged behind. Many coral reefs around the world are currently under significant threats. Capacity building, financial resources, and international collaboration should be ramped up, especially in developing countries and small islands where the socioeconomic threats are the largest.

## Author details

Pasita Chaijaroen
Vidyasirimedhi Institute of Science and Technology, Rayong, Thailand

*Address all correspondence to: pasita.c@vistec.ac.th

## References

[1] World Bank. Our Living Oceans. 2012. Available from: https://www.worldbank.org/en/news/feature/2012/05/30/infographic-our-living-oceans [Accessed: February 2, 2022]

[2] OECD. Sustainable Ocean for all2020. p. 173. DOI: 10.1787/bede6513-en Available from: https://www.oecd-ilibrary.org/content/publication/bede6513-en

[3] Moberg F, Folke C. Ecological goods and services of coral reef ecosystems. Ecological Economics. 1999;**29**(2):215-233

[4] Cesar H. Coral reefs: Their functions, threats and economic value. In: Collected Essays on the Economics of Coral Reefs. 2000. pp. 14-39

[5] Cesar H, Burke L, Pet-Soede L. The economics of worldwide coral reef degradation. 2003. Available from: https://www.wwf.or.jp/activities/lib/pdf_marine/coral-reef/cesardegradationreport100203.pdf

[6] Spalding MD, Dan Brumbaugh R, Landis E. Atlas of Ocean Wealth. Nature Conservancy; 2016

[7] Burke L et al. Reefs at risk revisited. World Resources Institute; 2011

[8] Ferrario F et al. The effectiveness of coral reefs for coastal hazard risk reduction and adaptation. Nature Communications. 2014;**5**(1):1-9

[9] UN Environment et al. The Coral Reef Economy: The Business Case for Investment in the Protection, Preservation and Enhancement of Coral Reef Health. 2018. Available from: https://wedocs.unep.org/handle/20.500.11822/26694;jsessionid=E749691E9B9D782E9C78A2AD9E325D63

[10] FAO. The State of World Fisheries and Aquaculture 2020. Sustainability in Action. Rome, Italy: FAO; 2020. 244 p

[11] Rogers A, Blanchard JL, Mumby PJ. Fisheries productivity under progressive coral reef degradation. Journal of Applied Ecology. 2018;**55**(3):1041-1049

[12] Pratchett MS, Hoey AS, Wilson SK. Reef degradation and the loss of critical ecosystem goods and services provided by coral reef fishes. Current Opinion in Environmental Sustainability. 2014;**7**:37-43

[13] Quentin R, Grafton. Adaptation to climate change in marine capture fisheries. Marine Policy. 2010;**34**(3):606-615

[14] Belhabib D, Lam VWY, Cheung WWL. Overview of west African fisheries under climate change: Impacts, vulnerabilities and adaptive responses of the artisanal and industrial sectors. Marine Policy. 2016;**71**:15-28

[15] Jara HJ et al. Current and future socio-ecological vulnerability and adaptation of artisanal fisheries communities in Peru, the case of the Huaura province. Marine Policy. 2020;**119**:104003

[16] Coulthard S. Adapting to environmental change in artisanal fisheries—Insights from a south Indian lagoon. Global Environmental Change. 2008;**18**(3):479-489

[17] Graham T, Idechong N, Sherwood K. The value of dive-tourism and the impacts of coral bleaching on diving in Palau. In: Coral Bleaching: Causes, Consequences and Response. Citeseer; 2001. pp. 59-71. Available from:

https://www.crc.uri.edu/download/COR_0011.pdf

[18] Cesar H et al. First evaluation of the impacts of the 1998 coral bleaching event to fisheries and tourism in the Philippines. In: Coral Bleaching: Causes, Consequences and Response. Citeseer; 2001. pp. 41-58

[19] Pratchett MS et al. Effects of climate-induced coral bleaching on coral-reef fishes—Ecological and economic consequences. In: Oceanography and Marine Biology. CRC Press; 2008. pp. 257-302. Available from: https://researchonline.jcu.edu.au/4901/https://www.taylorfrancis.com/chapters/edit/10.1201/9781420065756-8/effects-climate-induced-coral-bleaching-coral-reef-fishes-ecological-economic-consequences-morgan-pratchett-philip-munday-shaun-wilson-nicholas-graham-joshua-cinner-david-bellwood-geoffrey-jones-nicholas-polunin-tim-mcclanahan

[20] Cheablam O, Shrestha RP, Emphandhu D. Does coral bleaching impact tourists' revisitation? A case of mu Ko Surin marine National Park, Thailand. Journal of Food, Agriculture & Environment. 2013;**11**(3&4):2648-2654

[21] Samsuvan W, Yeemin T. Tourist perception of coral bleaching in the Andaman Sea. In: The 12th International Coral Reef Symposium Proceedings. Helsinki, Finland: WFL Publisher; Available from: https://www.wflpublisher.com/Issue/4/. 2012

[22] Prideaux B, Carmody J, Pabel A. Impacts of the 2016 and 2017 mass coral bleaching events on the great barrier reef tourism industry and tourism-dependent coastal communities of Queensland. In: Report to the Reef and Rainforest Research Centre Limited Cairns. Reef and Rainforest Research Centre Limited; 2017. Available from: https://www.rrrc.org.au/wp-content/uploads/2018/11/RRRC-Impacts-2016-17-Coral-Bleaching-on-GBR-Digital.pdf

[23] Doshi A et al. Loss of economic value from coral bleaching in SE Asia. In: The 12th International Coral Reef Symposium Proceedings. https://www.icrs2012.com/proceedings/manuscripts/ICRS2012_22D_1.pdf. 2012

[24] United Nations Ocean Conference. Factsheet: People and Oceans. 2017. Available from: https://www.un.org/sustainabledevelopment/wp-content/uploads/2017/05/Ocean-fact-sheet-package.pdf [Accessed: February 1, 2022]

[25] UNISDR. Global Assessment Report on Disaster Risk Reduction 2011: Revealing Risk, Redefining Development. 2011. Available from: https://www.preventionweb.net/english/hyogo/gar/2011/en/home/index.html [Accessed: February 1, 2022]

[26] Beck MW et al. The global flood protection savings provided by coral reefs. Nature Communications. 2018;**9**(1):1-9

[27] Cruz-Trinidad A et al. How much are the Bolinao-Anda coral reefs worth? Ocean & Coastal Management. 2011;**54**(9):696-705

[28] Hoegh-Guldberg O. Climate change, coral bleaching and the future of the world's coral reefs. Marine and Fresh water Research. 1999;**50**(8):839-866

[29] Wilkinson C, Hodgson G. Coral reefs and the 1997-1998 mass bleaching and mortality. Nature & Resources. 1999;**35**(2):16-25

[30] Goreau T et al. Conservation of coral reefs after the 1998 global bleaching

event. Conservation Biology. 2000; **14**(1):5-15

[31] BPS - Statistics Indonesia. Number of Fishing Boats by Province and Type of Boat for Marine Fisheries, 2000–2016. Cambridge, England: Cambridge University Press; 2019. Available from: https://www.bps.go.id/statictable/2014/01/10/1710/jumlah-perahu-kapal-menurut-provinsi-dan-jenis-perahu-kapal-2000-2015.html [Accessed: February 4, 2022].

[32] Chaijaroen P. Long-lasting income shocks and adaptations: Evidence from coral bleaching in Indonesia. Journal of Development Economics. 2019;**136**: 119-136

[33] Chaijaroen P. Tradeoffs between fertility and child development attributes: Evidence from coral bleaching in Indonesia. In: Environment and Development Economics. 2021. pp. 1-21

[34] Frankenberg E and Karoly LA. The 1993 Indonesian Family Life Survey: Overview and Field Report. Tech. Rep. 1995

[35] Frankenberg E. and Duncan T.. The Indonesian Family Life Survey (IFLS): Study Design and Results From Waves 1 and 2. Tech. rep. DRU-2238/1-NIA/NICHD. 2000

[36] Strauss J. et al. The Third Wave of the Indonesia Family Life Survey (IFLS3): Overview and Field Report. Tech. rep. WR-144/1-NIA/NICHD. 2004

[37] Strauss J. et al. The 4th Wave of the Indonesian Family Life Survey (IFLS4): Overview and Field Report. Tech. rep. WR-675/1-NIA/NICHD. 2009

[38] Becker GS, Gregg H, Lewis. On the interaction between the quantity and quality of children. Journal of Political Economy. 1973;**81**(2, Part 2):S279-S288

[39] Kim J, Prskawetz A. External shocks, household consumption and fertility in Indonesia. Population Research and Policy Review. 2010;**29**(4):503-526

[40] Sellers S, Gray C. Climate shocks constrain human fertility in Indonesia. World Development. 2019;**117**:357-369

[41] Currie J, Almond D. Human capital development before age five. In: Handbook of Labor Economics. Vol. 4. 2011. pp. 1315-1486

[42] Currie J, Vogl T. Early-life health and adult circumstance in developing countries. Annual Review of Economics. 2013;**5**(1):1-36

[43] Maccini S, Yang D. Under the weather: Health, schooling, and economic consequences of early-life rainfall. American Economic Review. 2009;**99**(3):1006-1026

[44] O'Leary BC et al. Effective coverage targets for ocean protection. Conservation Letters. 2016;**9**(6):398-404

[45] Behrman JR, Hoddinott J. Programme evaluation with unobserved heterogeneity and selective implementation: The Mexican PROGRESA impact on child nutrition. Oxford Bulletin of Economics and Statistics. 2005;**67**(4):547-569

[46] Kusuma D et al. The impact of household and community cash transfers on children's food consumption in Indonesia. Preventive Medicine. 2017; **100**:152-158

[47] Kusuma D et al. New evidence on the impact of large-scale conditional cash transfers on child vaccination rates: The case of a clustered-randomized trial in

Indonesia. World Development. 2017;**98**: 497-505

[48] Aizawa T. Joint impact of the conditional cash transfer on child nutritional status and household expenditure in Indonesia. Journal of Human Capital. 2020;**14**(1):122-164

[49] Cahyadi N et al. Cumulative impacts of conditional cash transfer programs: Experimental evidence from Indonesia. American Economic Journal: Economic Policy. 2020;**12**(4):88-110